TECHNICAL ANALYSIS AND FINANCIAL ASSET FORECASTING

From Simple Tools to Advanced Techniques

TECHNICAL ANALYSIS AND FINANCIAL ASSET FORECASTING

From Simple Tools to Advanced Techniques

Raymond Hon Fu Chan • Spike Tsz Ho Lee
The Chinese University of Hong Kong, Hong Kong

Wing-Keung Wong
Hong Kong Baptist University, Hong Kong

NEW JERSEY • LONDON • SINGAPORE • BEIJING • SHANGHAI • HONG KONG • TAIPEI • CHENNAI

Published by

World Scientific Publishing Co. Pte. Ltd.
5 Toh Tuck Link, Singapore 596224
USA office: 27 Warren Street, Suite 401-402, Hackensack, NJ 07601
UK office: 57 Shelton Street, Covent Garden, London WC2H 9HE

Library of Congress Cataloging-in-Publication Data
Chan, Raymond Hon Fu.
Technical analysis and financial asset forecasting : from simple tools to advanced techniques / by Raymond Hon Fu Chan (The Chinese University of Hong Kong, Hong Kong), Spike Tsz Ho Lee (The Chinese University of Hong Kong, Hong Kong), Wing-Keung Wong (Hong Kong Baptist University, Hong Kong).
pages cm
Includes bibliographical references and index.
ISBN 978-9814436243 (alk. paper)
1. Technical analysis (Investment analysis) 2. Stock price forecasting. I. Title.
HG4529.C4465 2014
332.63'2042--dc23
2014017657

British Library Cataloguing-in-Publication Data
A catalogue record for this book is available from the British Library.

In-house Editors: Sutha Surenddar/Alisha Nguyen

Typeset by Stallion Press
Email: enquiries@stallionpress.com

Printed in Singapore

To our families

Acknowledgments

We wish to express our sincere gratitude to Mr. Jason Sung-Hin Chan, Mr. Daniel Kei-Tsi Cheng, Mr. Teng Chen, and Mr. Tony Siu-Tung Ho for their comments and discussions. In addition, we are deeply grateful to Alisha Nguyen and Sutha Surenddar, editors of World Scientific Publishing. Last but not least, our thanks go to our families for their support throughout this book project.

Raymond H. Chan's research is supported by HKRGC Grant CUHK 400412.

Preface

Technical analysis is more than just drawing lines on a stock price chart to look for market entires. This book presents an expansive collection of tools in technical analysis, from classic concepts like chart patterns, simple quantifiable methods like moving averages, to advanced mathematical techniques like wavelets and empirical mode decomposition.

The major targeted audience is university undergraduates with sufficient mathematical and statistical knowledge. Some advanced mathematical concepts such as wavelets and Hilbert transform would appear but they would not be a hindrance. In the end, it is all about implementation of all these technical methods on a computer. In particular, we mainly make use of the MATLAB environment and illustrate the examples using various functions provided by the MATLAB toolboxes.

Some of the materials in this book are research related, but we skip the details in the context in order to make the book more approachable to a general audience. Postgraduates and researchers in the fields of technical analysis and financial markets are referred to the journal papers for a more detailed analysis.

The book first discusses some traditional tools in technical analysis, such as trend line and trend channel, which can be classified as chart pattern reading. To visualize these subjective methods, we produce a number of figures using the recent data of companies listed on the Hong Kong Stock Exchange. The book then moves on to study basic technical analysis tools like the moving averages. We consider these quantifiable methods and lay out their properties from the perspective of filtering theory. To measure the profitability of these techniques, we perform statistical tests on some recent data like the Hang Seng Index from the period 2007 to 2011. The last part of the book introduces other existing technical indicators like Gann's line and Bollinger bands. We also provide the mathematical expressions for some technical indicators that are usually described by words only.

The book aims to give readers a roundup of the various methods used in technical analysis, and at the same time, elaborate on the mathematical aspects of several technical filtering tools.

Raymond H. Chan
Spike T. Lee
Wing-Keung Wong

Contents

List of Figures

List of Tables

PART 1

Classical Technical Analysis

Chapter 1

Introduction to Technical Analysis

1.1 What Is Technical Analysis?

Technical analysis (TA) is a broad area of study. Technical analysis is the tracking and prediction of asset price movements using charts and graphs in combination with various mathematical and statistical methods. More precisely, it is the quantitative criteria for predicting the relative strength of buying and selling forces within the market to determine what to buy, what to sell, and when to execute trades. Technical analysis should be distinguished from *fundamental analysis*, its counterpart in the world of price forecasting and market action, which involves the evaluation of elements that measure, reflect, or influence the supply of and the demand for a firm's goods or services. While technical analysis was once considered as nothing more than the art of "chart-reading", it is now an accepted field of study in the analysis of the stock market. A great number of mathematical and statistical indicators are increasingly being introduced and integrated with prevailing practices, thus building technical analysis into a scientific and sophisticated system of tools and techniques to monitor and estimate trends in the stock market, both macro and micro.

Price is of the utmost importance in applying technical analysis to the stock market. Traders and investors use market filters or indicators to capture a watchlist of shares (see Figure 1.1). The diligent technical analyst will concentrate primarily on the price movement on an asset. Such an approach is based on the premise that market action discounts everything; whatever factor can possibly affect the market price of a stock, be it fundamental, political, psychological, or others, is already reflected in the price of that stock. The trading volume, if need be, is used as a secondary check in the

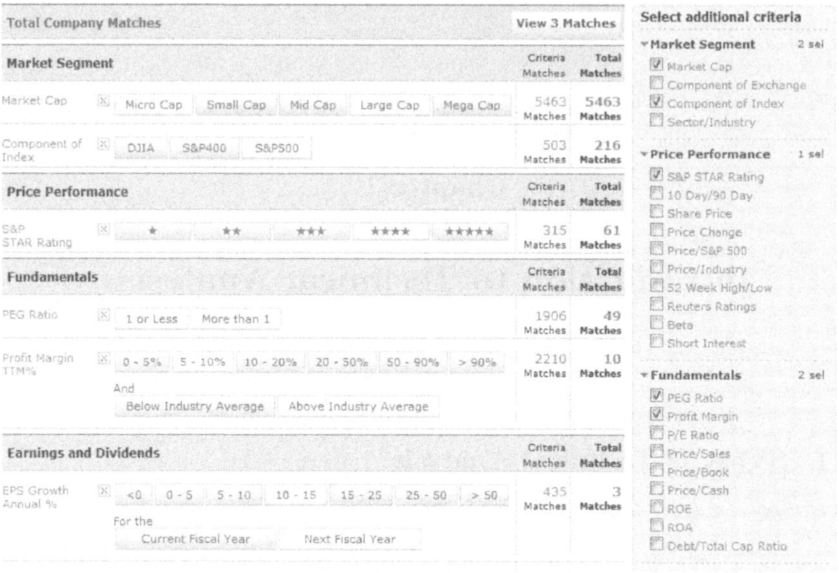

Figure 1.1: Stock screener on online discount brokerage Scottrade

analysis. The technical analyst will then identify any price trends with the aid of a chart, an essential tool in technical analysis.

A *trend* is a price movement in motion that continues in the same direction until it reverses in an opposite direction due to changes in force. Since the price movements of a stock over a reasonable period of time are plotted in the form of a chart, certain patterns will be formed from the past performance of the stock prices, and trends are easily spotted. If such a chart pattern shows a *trend line*, straight lines drawn through the troughs or the pinnacles of an asset price line within a specific period (see Figure 1.2), it is considered a clear trend, which the technical analyst will take into account in his analysis before predicting the timing of a reversal pattern. Although spotting a reversal trend is a challenging part in technical analysis, the breaking of an important trend line is usually the first sign of *reversal.* With these signals, the technical analyst decides on when and whether to buy or sell a stock.

1.2 Origin of Technical Analysis and Its Development

The application of technical analysis as a trading method could be traced back to the trading of Japanese rice on the Dojima Rice Exchange in Osaka as early as late 1600s. One approach, known as "Japanese candlesticks"

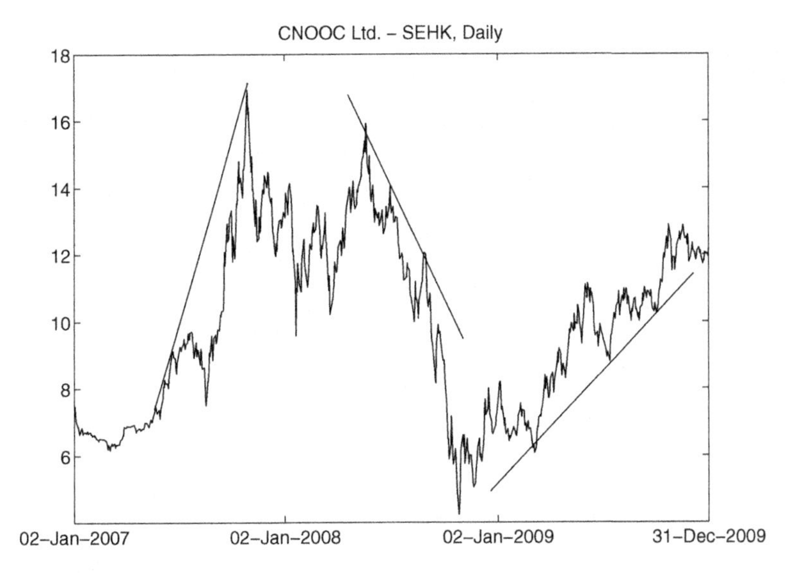

Figure 1.2: Clear trend lines on typical stock price movement

(a chart showing "candles" formed by the opening, closing, high and low prices of a commodity), was documented by a certain Munehisa Homma, born in 1724. However, this trading method was not widely adapted until technical analysis became popularized in the late 1980s. In the West, technical analysis started in the early 1900s with point and figure charting and the Dow theory [Schannep (2008)], then evolved into chartism in the early twentieth century. One main criticism that slows its mainstream acceptance is that chartism is inherently subjective. To counter this, technical analysis later developed into precise mechanical trading rules as well as complicated trading systems with multiple data input. Since mathematical formulas are used to generate signals under these sophisticated trading rules and systems, subjectivity is eliminated from the interpretation of the signals. This development is aided by the introduction of computers which take the tedium out of complex mathematical manipulations.

Nowadays, technical analysis covers the more exotic methods, like the use of astronomical cycle, 28-day trading cycle (the Lunar cycle), 10.5-month futures cycle, January effect, 4-year cycle (the Kitchin cycle [Kitchin (1923)] and the presidential election cycle [Wong and McAleer (2009)]), 9.2-year cycle (the Juglar cycle), 54-year cycle (the Kondratieff cycle [Kondratiev (1925)]), Elliott wave [Elliott (1938); Frost and Prechter (2005)], as well as Gann's trading methods [Gann (1935); Reddy (2012)].

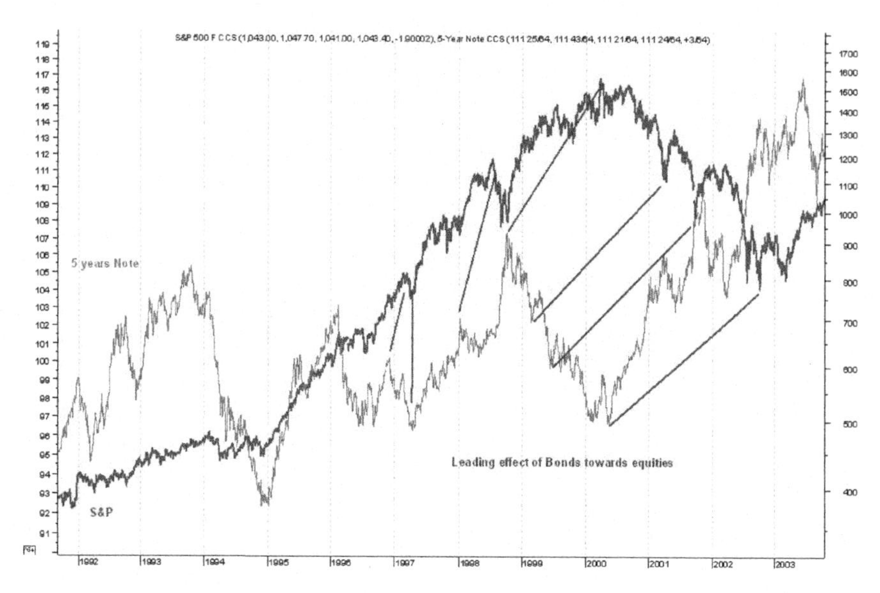

Figure 1.3: The "leading effect" of 5-year Treasury notes over the S&P 500

Meanwhile, high-capacity computers initiate a family of modern-day indicators that combine traditional price and volume data with other factors such as economic data. The relationship between commodities, US dollar, world currencies, bonds, interest rates and equities are often explored to develop innovative indicators. A typical trading signal could simply be the combination of gross domestic product (GDP) growth rate, interest rate and price-to-earnings (P/E) ratios, with asset price and volume data. One classic example is that movements in bond prices are often a good estimator of movement in equity indices: the prices of 5-year Treasury notes are shown to have successfully predicted the S&P 500 with considerable foresight, known as the "leading effect" (see Figure 1.3). In another example, since a clear inverse relationship exists between the US dollar and the Commodity Research Bureau (CRB) Index, and because the relationship between the prices of 5-year Treasury notes and the CRB Index is positive (see Figures 1.4 and 1.5, respectively), the CRB Index becomes a reliable technical indicator of the price movement of assets correlated with it. Additionally, other methods such as artificial neural network, genetic algorithms, chaos theory, and fuzzy logic are actively employed to meet the needs from the financial industries. These are convincing proofs that technical analysis has been constantly evolving and is no longer confined to simply chart pattern reading.

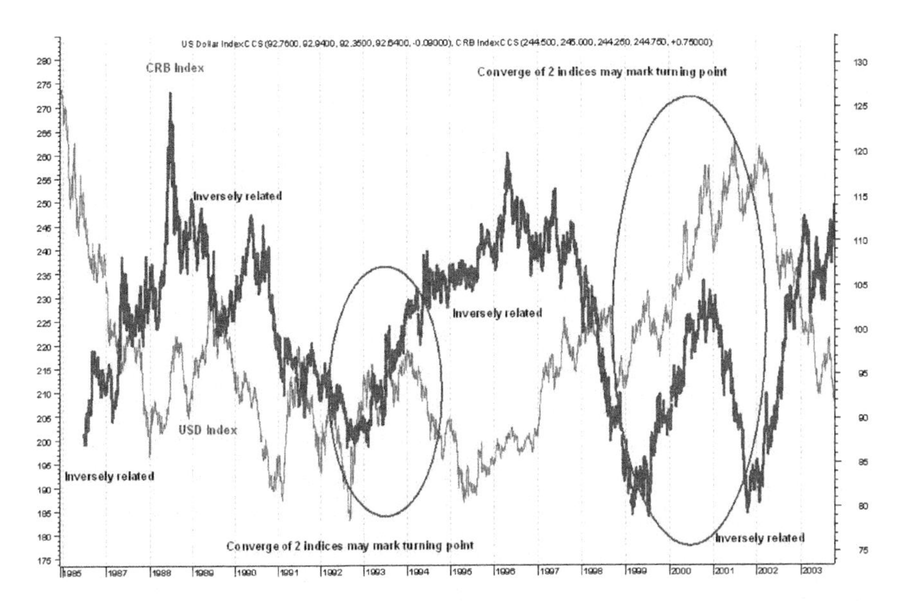

Figure 1.4: Inverse relationship between the US dollar and the CRB Index

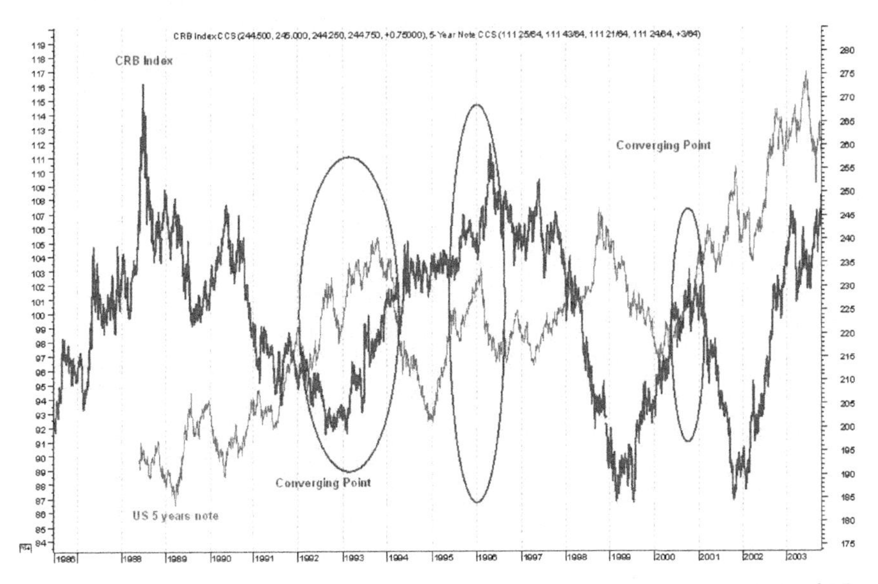

Figure 1.5: Positive relationship between the 5-year Treasury notes and the CRB Index

1.3 Application of Technical Analysis in Stock Markets

The application of technical tools for market timing decisions has been the subject of much discussion. Some academics have questioned the usefulness of technical analysis, arguing that traders using technical analysis indicators usually cannot produce better returns than those who apply the *buy-and-hold strategy* using fundamental indicators. For example, studies done on the rules that dictate buy and sell signals, known as *filter rules*, conclude that traders on the New York Stock Exchange (NYSE) do not generate significantly superior returns using filter rules compared to investors using a buy-and-hold strategy. In particular, returns could turn out to be negative if transaction cost is considered [Fama and Blume (1966); Jensen and Bennington (1970); Ball (1978)]. Furthermore, it has been suggested that an accuracy of 70% or more is required for any market timing strategy to produce returns in excess of the buy-and-hold strategy [Sharpe (1975)]. These studies, however, judge the value of market timing strategies solely by comparing its returns with the buy-and-hold strategy returns [Samuelson (1989)].

Practitioners, on the other hand, have frequently claimed that they can achieve superior performance with sound market timing skills. By using a proper system of indicators, either technical or economic, traders claim to be able to generate enormous returns. In fact, it is noted in Allen and Taylor (1989) that "some chartists' forecasts were remarkably good and could not be improved upon by a range of alternative forecasting procedures". Indeed, in some financial markets, traders have been known to generate large profits consistently over a relatively long period of time (Figure 1.6 shows the successful application of technical analysis signals in trading). Thus the argument puts forth, that non-fundamentalist traders would be quickly driven out of business by fundamental buy-and-hold investors, continues to remain under close scrutiny [Friedman (1953); Fama (1970)].

Various reports reveal that practitioners do turn to technical analysis. From Allen and Taylor (1989, 1992), about 90% of chief dealers make use of technical indicators to estimate the expected future price movements. Another 60% of them feel that technical analysis is at least as important as fundamentals. Technical analysis comes right after fundamental analysis among investment managers, while portfolio analysis is the lowest rated [Carter and Van Auken (1990)].

Including technical analysis in financial market forecasting is not unusual among professional investors. A survey by Euromoney reveals that

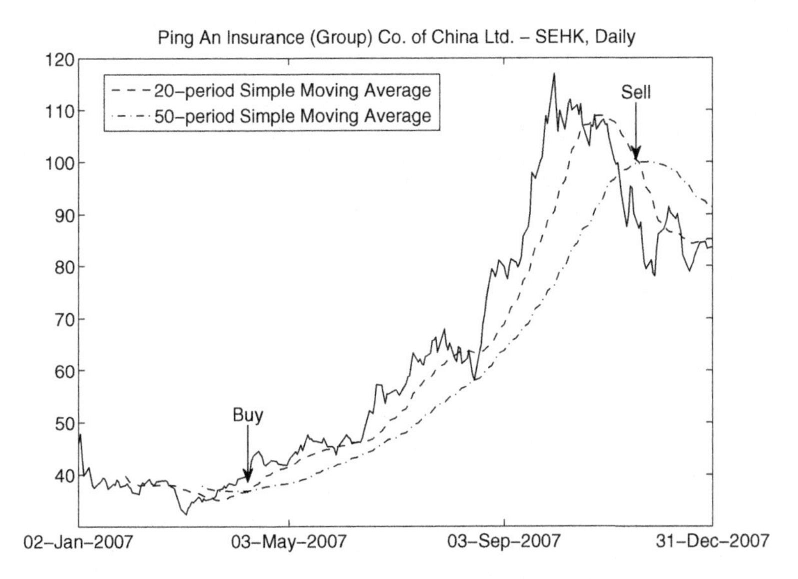

Figure 1.6: Buy and sell signals on the China Ping an Insurance

in the 80s, there exists a transition from fundamental analysis to technical analysis [Frankel and Froot (1990a)]. Lui and Mole (1998) find that "more than 85% of respondents rely on both fundamental and technical analysis for predicting future rate movements at different time horizons". Technical analysis is "considered slightly more useful in forecasting trends than fundamental analysis, but significantly more useful in predicting turning points". Traders in Hong Kong, Japan and Singapore also rely on technical analysis heavily in foreign exchange markets [Cheung and Wong (1999)].

On the other hand, detailed and updated technical analysis information can be obtained conveniently from real-time financial data providers like Reuters and Bloomberg. This is another indication that technical analysis is highly demanded, which also results in competition among financial information services to offer the latest technical analysis tools.

1.4 Basic Principle of Technical Analysis

The basic principle of technical analysis is to identify and go along with the trend. This assumes that there is an uneven distribution of information — that "smart money" acts on information before it becomes public. Thus, asset prices would reflect the information available to "smart money". This is consistent with the idea of costly information, where competitive incentives for traders to seek information results in an equilibrium where information is passed along through prices [Grossman and Stiglitz (1976); Grossman (1976)].

We can conclude that because of its popularity among the ordinary as well as professional investors, the information extracted by technical analysis will be valuable. Technical indicators are inclined to give signals in the prevailing direction of a trend, while the root of the trend can be any sudden news or fundamental aspects. In this regard, trends are strengthened and supported by many investors who use the same technical indicators. As a result, consequent actions from these technical indicators will provoke a self-fulfilling phenomenon regardless of the cause of the trend, and this phenomenon is known as the "sunspot theory" [Curcio and Goodhart (1991)]. This self-fulfilling nature of technical analysis will lead to an enhanced accuracy of technical indicators. However, feedback theory suggests that if trends develop solely due to the propagation of technical indicators, speculative bubbles will emerge within the market. This "herd" tendency is particularly apparent for short-term traders [Froot *et al.* (1992)]. In a way, going with the crowd does increase the probability of profit. Rational traders might follow the market herd tendency to be entitled to more profitable returns. This could be the reason for studies reporting positive auto-correlation for weekly returns [De Long *et al.* (1990)].

Weekly returns on portfolios of NYSE stocks group according to size demonstrated reliable positive auto-correlation. More importantly, the correlation is more pronounced for portfolios of small stocks [Conrad and Kaul (1988)]. Excess bond and stock market volatility is often caused by irrational investor behavior. Evidence suggests that the overpricing of the US dollar in the 1980s with respect to the underlying economic fundamentals [Frankel and Froot (1986, 1990b)] and the October 1987 world-wide stock market crash (see Figure 1.7) could be due largely to the influence of technical analysis [Shiller (1984, 1987)].

1.5 Usefulness of Technical Analysis in Stock Markets

More recently, however, the importance and usefulness of market timing studies are being discussed. Studies indicate that stock price movements can be explained by a mean reverting model, with auto-correlation of returns becoming strongly negative over a 3- to 5-year horizon [Fama and French (1988)]. Stocks that have been extreme winners/losers for a certain period tend to have weaker/stronger returns relative to the market during the following years [DeBondt and Thaler (1985, 1987)]. New experiments on

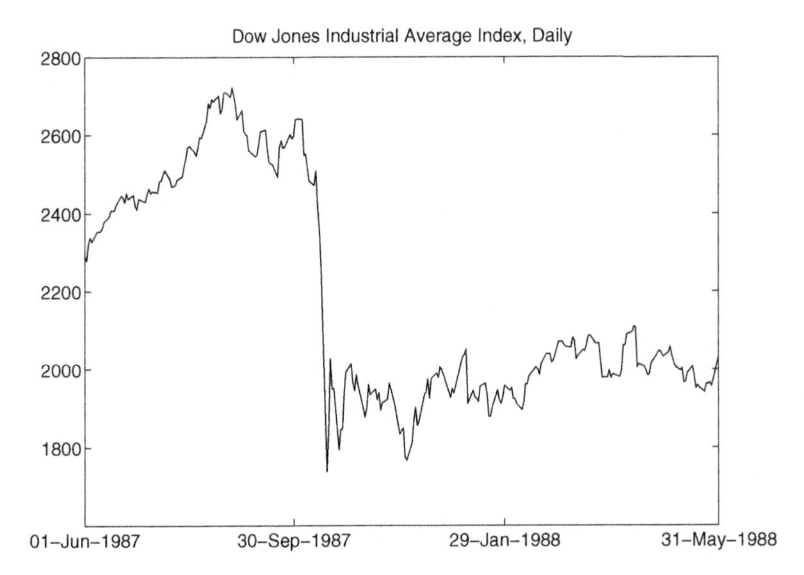

Figure 1.7: The October 1987 stock market crash

technical analysis now refute original conclusions, saying that there is no need for predictive accuracy to be as high as 70% for the gains to be large. In Sy (1990), it is shown that market timing will be more and more fruitful when the difference in cash returns and stock returns decreases and market volatility increases.

Small set of filter rules are also found to be profitable when applied to daily exchange rates in currency markets, after taking into account the interest expense, interest income and transaction costs [Sweeney (1986)]. A comprehensive study using several technical indicators, including the popular method of using the crossover between a short term and a long term moving average as trading signals, revealed that most "simple technical trading rules have very often led to profits that are highly unusual" [Levich and Thomas (1993)], even after adjusting for interest expense and transaction costs [Schulmeister (1988)]. Stock returns could be predicted empirically based on rational aggregate output [Balvers *et al.* (1990)]. Fundamental or economic data like business conditions, dividend yields, and P/E ratios can forecast stock returns to a decent level as well [Breen *et al.* (1989); Campbell (1987); Campbell and Shiller (1988a,b); Cochrane (1991); Fama and French (1989); Renshaw (1993)].

Locally, Wong and Wong (1993) uses the Hong Kong stock market index to show that technical indicators give "overwhelming" signals in stock price dynamics and that trading rules based on technical trend signals generate significant returns that outshine the buy-and-hold strategy. Wong also finds "significant positive (negative) nominal and excess returns in the period after the signals change to bullish (or bearish)" [Wong (1995)]. This suggests that technical analysis does provide valuable information regarding the market, and there are to a certain degree "herd" tendencies within the market. Traders have incentive to employ trend-following indicators in the Hong Kong market as there is "strong evidence that some rule-based portfolios have a consistently and significantly superior market timing ability regardless of investment horizons" [Wong (1997)].

Portfolio performance could be improved by market timing [Vandell and Stevens (1989)]. Apart from higher rates of return and terminal wealth, market timing also impacts the volatility of portfolios when employed, as "... lost opportunities for gains in good markets were more than made up when dramatic downturns occurred". By applying simply market timing skills, portfolios as a whole do have superior performance than the buy-and-hold portfolio.

As stated earlier, traditional technical analysis can be combined with fundamental or economical data to create insightful indicators. One such indicator is the *standardized yield differential* (SYD) (formerly known as SRP — standardized risk premium) [Wong (1993, 1994)]. It utilizes the difference between the earnings-to-price yield (E/P ratio, the reciprocal of the more common P/E ratio) and the bond yield or the interbank interest rate. Examples therein focus solely on the Singapore market and are more descriptive rather than analytical in nature. Studies show that the use of the SYD is a promising technique for stock market investment. Applying linear regression to the SYD indicator, one observes a "significant relationship between the aggregate risk-premium and share prices" [Ariff and Wong (1996)].

1.6 Filtering Tools in Technical Analysis

In 1970, Fama (1970) developed the three common forms of the efficient-market hypothesis (EMH): weak-form, semi-strong form and strong-form. The weak-form EMH states that the present-day stock prices or bond prices already reflect all publicly known information from the past. A market with weak-form efficiency makes it impossible for investors to predict recurring price patterns and make excess returns by using trading strategies derived from historical data.

Traders, however, think that some forms of market inefficiency exist and it is possible to use price patterns and historical data, i.e., technical analysis, to beat the buy-and-hold strategy. The classic technical analysis tool is chart pattern reading. Technical analysts believe that a chart pattern provides broad and helpful clues on asset price movements. Investors then act according to what they see in the chart patterns. However, chart pattern reading cannot be quantified mathematically and the reading of chart patterns can be very subjective. Therefore, different people reading the same chart could have different interpretations.

The subjectivity of chart pattern reading leads people to examine technical indicators. Indicators are essentially generated by mathematical filtering tools. They are more objective and not as open to interpretation as chart patterns alone. An extensive review of this line of research can be found in Park and Irwin (2007). A popular and widely-used trending indicator is called the moving average [Pring (2002); Mak (2006)]. It is simply the average of a fixed number of previous stock prices, where the number is the duration of the moving average. Traders then determine whether to enter the market at the crossovers of two moving averages with different durations. Fundamentally, a moving average is a kind of convolution. By studying its frequency response, we can see that the difference of two moving averages is equivalent to a band-pass filter, which has the ability to detect market entries and exits.

A seminal paper by Brock *et al.* (1992) finds strong support for the effectiveness of the moving average and the trading range break rules in the Dow Jones Industrial Average Index. Later on, Bessembinder and Chan (1995) extend Brock *et al.*'s study [Brock *et al.* (1992)] in six Asian markets. They find that the trading rules are successful in the emerging markets of Malaysia, Thailand and Taiwan but are less satisfactory in more developed markets, such as Hong Kong and Japan. Mills (1997) tests simple technical trading rules on the London Stock Exchange FT30 Index from 1935 to 1994. The returns are significantly better than the buy-and-hold strategy at least up to the early 1980s, i.e., when the market was effectively driftless.

More studies on simple technical rules have been carried out in the 2000s. Kwon and Kish (2002) work on the NSYE. They find out that the profits are significantly weakened over time, implying that the market is becoming more efficient. On the other hand, Tian *et al.* (2002) show the profitability in the Chinese stock market even in the presence of trading cost. Wong *et al.* (2003) use the moving average and the relative strength index to test Singapore data and obtain significantly positive returns. In Wong *et al.* (2005), Wong *et al.* use simple trading rules like moving average to argue that technical analysis indeed outperforms the buy-and-hold strategy in China, Hong Kong, and Taiwan. In Vasiliou *et al.* (2006),

Vasiliou *et al.* investigate different technical trading rules in Athens' stock market covering from January 1990 to December 2004. The results therein back up the profitability of technical strategies. A comprehensive review paper by Park and Irwin (2007) on technical analysis compares 95 modern studies on the usefulness of technical analysis and observes that 56 of them find positive results, 20 of them obtain negative results, and 19 of them indicate mixed results.

More new papers discuss the profitability of technical trading rules after Park and Irwin's review paper in 2007. Kung and Wong (2009) show that the Taiwan stock market has been greatly strengthened by the liberalization measures implemented over the last 20 years as their results show that the predictability of simple technical rules is deteriorating over the period. In Kung *et al.* (2010), Kung *et al.* use moving average rules and trading range breakout rules to show that the Indonesian market has become significantly more efficient over time. Cheung *et al.* (2011) study the profitability of simple moving average and trading range break in the Hong Kong stock market and examine the impact of market integration. Mitra (2011) performs the simple moving average rules in the Indian stock market using four stock index series and find that the Indian market is not efficient in the weak-form.

Meanwhile, many researchers have found evidence for the possibility that the financial markets may be nonlinear dynamical systems with important implications in the EMH. Several researchers, by using different statistical tests, have discovered evidence of non-independently and identically distributed (i.i.d.) behavior and nonlinear dependence in financial time series. On the other hand, moving averages can be categorized as linear filters, as they convert a time series into another time series by a linear transformation. A recent method known as the empirical mode decomposition (EMD) performs the conversion in a nonlinear manner, and it is also considered in financial market trading [Huang *et al.* (1998, 2003)]. Essentially, the EMD is an algorithm which decomposes a data series into a set of intrinsic mode functions. This kind of function has instantaneous frequency as a function of time, which depicts real-world behaviors more accurately.

1.7 Technical Analysis Strategies for Stock Market Investment

Reading and understanding this book will provide the basis for using technical analysis as a strategic tool for investing in the stock market.

Part 1 explains the fundamentals of technical analysis, and appreciates the tools and charts which are the stock-in-trade of the technical analyst. Chapter 1 has just given an overview of technical analysis, including the historical background of the study of technical analysis and the critiques of technical analysis. Chapter 2 introduces the primary tools for technical analysis, such as the various types of charts and how they are formed. Chapters 3 and 4 discuss chart pattern reading, important chart patterns and identification of market trends.

Part 2 covers the filtering theory in technical analysis. We study linear time-invariant systems and how to obtain a filter's behavior via its frequency response in Chapter 5. In Chapter 6, we study the frequency responses of several useful technical indicators such as the simple moving average, the exponential moving average, and the moving average convergence–divergence (MACD) indicator. In Chapter 7, we explore the family of momentum indicators and measure the usefulness of these commonly used technical tools with statistical tests. In Chapter 8, we move on to wavelet filters and create the corresponding momentum indicators for market entries and exits. In Chapter 9, we take a look at the young and refreshing tool in time series analysis, namely the empirical mode decomposition (EMD), and combine it with momentum indicators to make trading decisions.

In Part 3, we study some other existing technical analysis approaches. In Chapter 10, we unveil the mystery of Gann's theory which uses a unique "geometric" approach in breaking down the trend structure into segments. Then, we examine the effectiveness of Gann's trend lines and retracements in determining stock price movements. In Chapter 11, the moving average concept is extended as we investigate how Bollinger bands can be applied to the stock market in determining the support and resistant lines for stock prices. In Chapter 12, we introduce other existing technical trading methods, such as the standardized yield differential (SYD), which aims at integrating fundamental and economic variables into technical analysis, and scrutinize its effectiveness as a technical indicator for stock market investment. The relative strength index (RSI), currently a popular technical tool among technical analysts in a non-trending market situation, is also discussed.

Chapter 2

Primary Tools for Technical Analysis

As mentioned in Chapter 1, technical analysis involves the study of market action primarily through the use of charts to forecast future price trends. Hence, *price chart* becomes the most fundamental tool for technical analysis.

Plotting of charts is basically a process of recording the price and volume changes at different time intervals, e.g., hourly, daily, weekly or monthly. The choice of a suitable time period is dependent on the intended purpose defined by the technical analyst or investor. Generally, short-term traders use the daily and weekly charts, or even the hour, minutes or number of ticks chart for intraday trading, while long term traders would work from the weekly or monthly charts.

2.1 Types of Chart

The types of chart that are commonly used in technical analysis are:

(1) Bar chart
(2) Line chart
(3) Point and figure chart
(4) Candlestick chart

Bar Chart

Bar chart is acknowledged as the type of chart most widely used in technical analysis. In plotting price movement, four sets of prices are normally reflected by a vertical bar: the opening, high, low and closing prices. The length of the vertical bar is the difference between high and low prices

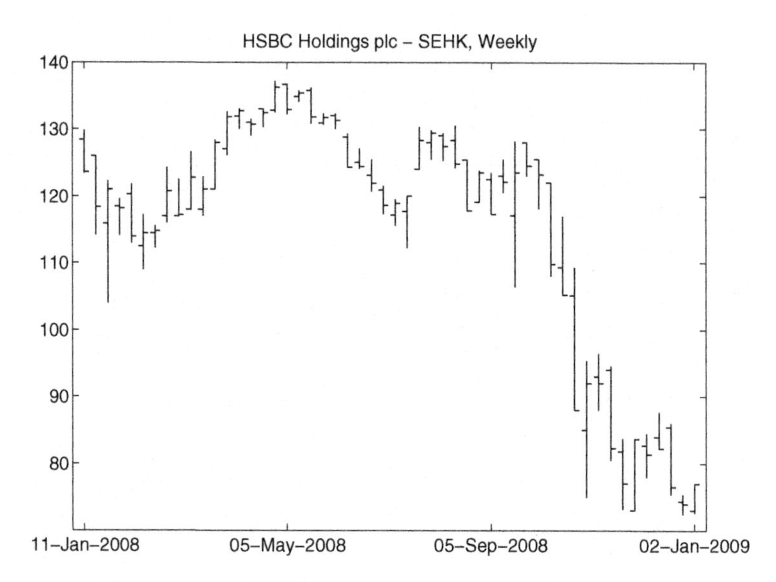

Figure 2.1: An example of a bar chart for HSBC Holdings

reached during intraday trading, while ticks on the bar to the left and right indicate the opening price and closing price, respectively.

Bar chart is plotted on X–Y plane. The X-axis represents the time interval, while the Y-axis the price interval. Volume of the asset traded during the time interval, if needed, are plotted separately below the bottom of the bar chart. Figure 2.1 shows an example of bar chart for HSBC Holdings. This can be done by using the `highlow` command in MATLAB. The presentation of bar chart enables traders to see the price movement within a fixed time period at one glance. It is a simple and easily understood type of chart for general investors.

Line Chart

Line chart is another commonly used chart in technical analysis. It resembles the bar chart except that the line chart shows only one set of price: the closing price of an asset for the day, plotted as a point on the X–Y plane. The line chart is drawn by connecting each of the asset's closing prices to its closing price on the previous day by a straight line. See Figure 2.2 for an example of line chart for HSBC Holdings.

Line chart presents price movements with a clean and simple look. This allows technical analysts to spot trends more easily. This type of chart is usually good for longer-term price trend analysis.

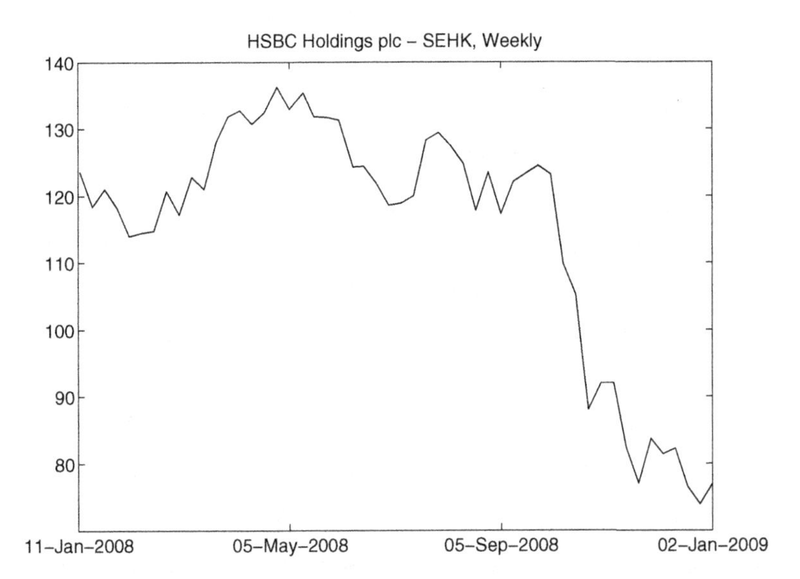

Figure 2.2: An example of a line chart for HSBC Holdings

Point and Figure Chart

As with all types of price charts, the vertical Y-axis on the *point and figure chart* is the price scale. However, unlike on the bar chart or line chart, the horizontal X-axis of the point and figure chart does not show a linear representation of time. Asset prices on the point and figure chart are recorded only when the price change reaches a certain minimum or maximum level set by the chartist. Accordingly, during time intervals with no significant price changes, nothing will be plotted. In other words, any new columns on the point and figure chart will indicate a reversal in asset prices. Instead of using simple lines as to represent price movements, point and figure charts use "X" and "O" to indicate upward and downward price movements, respectively. See Figure 2.3 for a point and figure chart for HSBC Holdings. In MATLAB, this can be charted by using the command `pointfig`.

Practically, it is not easy to plot an accurate point and figure chart; one will need to continuously monitor real-time asset prices or use detailed intraday data. The first data point on the chart is the difference between the high and low prices reached by the asset during intraday trading. Before the point and figure chart can be plotted, it is imperative to determine the asset price trend from the previous day: an uptrend occurs if closing price is

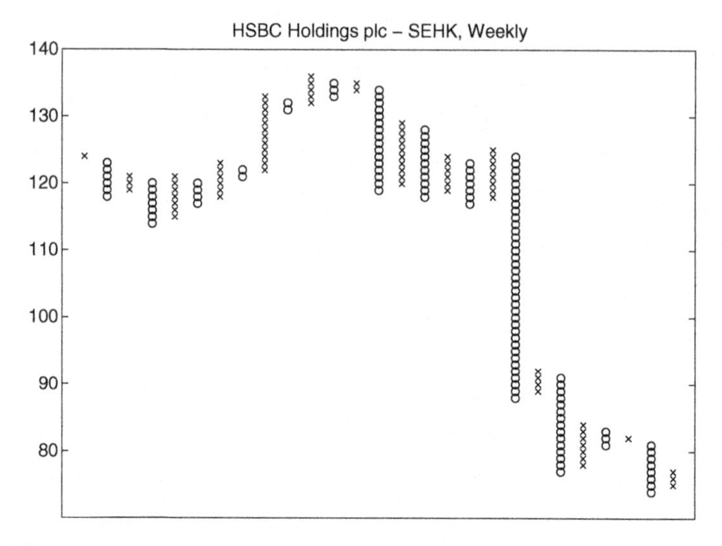

Figure 2.3: An example of a point and figure chart for HSBC Holdings

above opening price on the previous day; a downtrend occurs if the situation is reversed. One assumes that the current asset price trend follows the trend from the previous day, e.g., if one observes an uptrend on the previous day, one will mark "X" onto the uptrend column should the current day high exceed the previous high. If current day price movement fails to break the previous high, one then looks for the current low and check if the reversal amount is touched. If a reversal occurs, then one will begin to plot "O" in the subsequent column.

The price plotted in this type of chart is usually the closing price following an uptrend or a downtrend. Thus, it is seen as having the advantage of smoothening out minor fluctuations in asset prices before the close of a specific trading period, e.g., intraday trading. Note that instead of absolute price changes, some traders prefer to use relative percentage changes. Point and figure charts also require less plotting space, since the asset price data from time periods that follow the same trend are compressed into single columns.

However, due to its complex manner of presentation, the point and figure chart is not as popular among general investors who have grown accustomed to the familiar bar or line charts. Yet this charting style is popular amongst advanced traders as point and figure chart infers a very disciplinary way of trading: buy asset once price breaks its previous uptrend's high, sell asset when price breaks its previous downtrend's low. This is a concept that is very similar to trading with *trend lines*, a term discussed in Chapter 3.

Candlestick Chart

This charting method was first introduced in Japan, and has become the origin of technical analysis development over the years [Nison (1994, 2001)].

Similar to the bar chart, the *candlestick chart* also reflects the opening, high, low, and closing prices. The main visual difference is that on the candlestick chart, ticks on the left and right are extended across the vertical bar and joined to form a rectangle, also known as the real body, i.e., the opening and closing asset prices are represented by the extreme ends of the real body. If the price of the asset closes above its opening price, the real body is white (unshaded); if the opposite is true, the real body is coloured (shaded). The length of the bar that extends above and below the real body is known as the upper and lower shadows, respectively. An example of candlestick chart for HSBC Holdings is shown in Figure 2.4. This can be plotted by using the command `candle` in MATLAB.

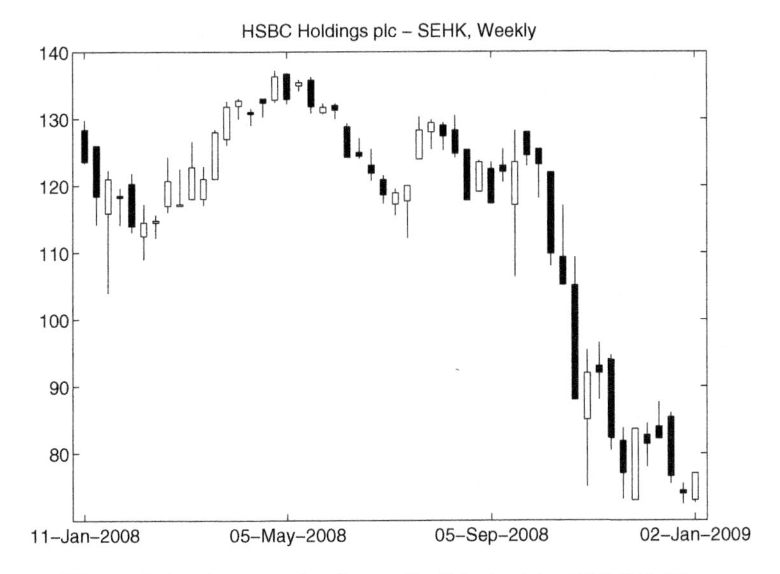

Figure 2.4: An example of a candlestick chart for HSBC Holdings

In a special situation where the opening price is equal to the closing price, the candlestick chart will look identical to an ordinary bar chart, also known as a *doji* signal and resembles a cross or star. Alternatively, if the opening price is also the low price and closing price happens to be the high price reached during intraday trading, this phenomenon is known as the *white marubozu* signal and resembles a white rectangle; if the situation is reversed, we would observe the *black marubozu* signal, which resembles a black rectangle.

With the candlestick chart, investors can easily determine whether bull or bear trends are dominant on particular trading periods simply from the colour, the height of the real body and the length of the shadows on the candlestick. A cursory analysis of asset price trends relies on the observation of the real body: the greater the length of the real body, i.e., the larger the difference between the opening the closing prices, the greater the strength of the prevailing trend. For example, if the bull trend is significantly stronger than the bear trend, the bull trend would be able to push asset prices much higher than the weak bear trend can counteract, leading to a sizable disparity between opening and closing asset prices. In the event that the bull and bear trends are equally strong, i.e., the market is indecisive, the candlestick will have no real body and doji will appear.

The candlestick chart is quite widely used by short-term traders, especially for futures trading. However, investors believe that the strength of the candlestick chart often lies in identifying reversal rather than continuation trends, as a majority of the 60 well-known candlestick chart indicators and single candle/group candle patterns are used to spot changes in the direction of asset price movements.

2.2 Price Chart Scaling

As seen from the figures illustrated, asset prices are traditionally plotted on the vertical Y-axis. Depending on the typical price movements of an asset, technical analysts may apply different scales to the chart with the intention of easily spotting trends with the naked eye. Generally, only two different scales are used to present price movements, namely the *arithmetic scale* and the *logarithmic scale.*

Arithmetic Scale

For the arithmetic scale, each unit of absolute price change is shown in equal distance on the vertical Y-axis. Arithmetic scale is usually used in

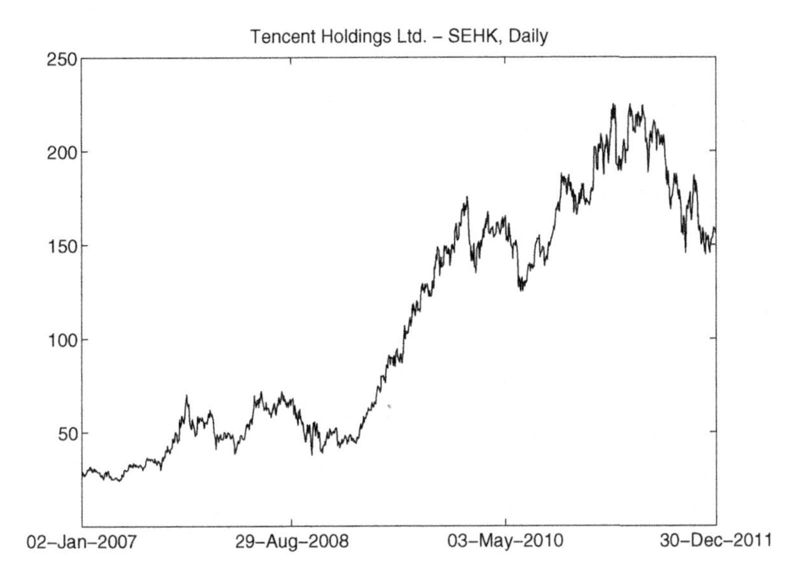

Figure 2.5: Price chart of Tencent Holdings Limited in arithmetic scale

commodities and interest rate futures due to their low trading volatility. Figure 2.5 shows an arithmetic scale chart for Tencent Holdings Limited.

Logarithmic Scale

It is well known that the first difference of logarithmic function gives the percentage change. Given today's closing asset price as x and tomorrow's one as y, we have the following approximation relation by Taylor series:

$$\log(y) - \log(x) = \log\left(\frac{y}{x}\right) = \log\left(1 + \frac{y-x}{x}\right) \approx \frac{y-x}{x}.$$

Hence, for the logarithmic scale (ratio scale), each similar percentage change in price is plotted in equal distance on the vertical Y-axis. Apart from percentage representation, the logarithmic scale also serves the purpose of reducing the price sizes. Logarithmic scale is commonly used in equities and FOREX because price movements are usually more extreme. Figure 2.6 is a logarithmic scale chart for Tencent Holdings Limited.

Charts with logarithmic scale are advantageous for long term trend analysis. Although these charts are not easily produced via manual plotting, they are prevalent in popular technical analysis computer software, even allowing for quick comparisons and switch-around between the arithmetic and logarithmic scales where necessary.

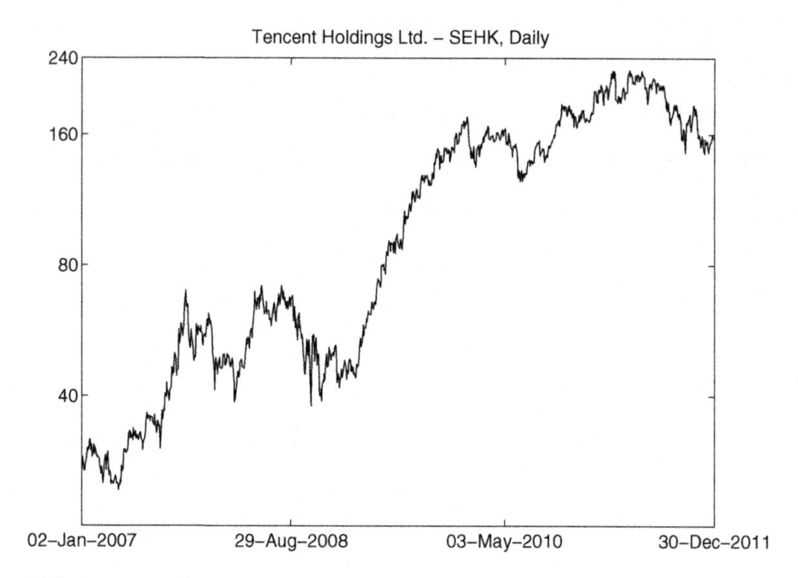

Figure 2.6: Price chart of Tencent Holdings Limited in logarithmic scale

2.3 Summary

A price chart is only a record of asset prices over an interval of time. It has little value or meaning in itself. How useful a price chart is depends largely on the interpretation of the technical analyst, trader or the investor. The basis of technical analysis begins with locating chart patterns formed by price movements and identifying their meanings. Nowadays, there are already an arsenal of chart patterns [Bulkowski (2005)], and traditional techniques like Japanese candlesticks [Nison (1994, 2001)] and ichimoku cloud [Linton (2010)] are also actively applied. With the advancement of technology, chart reading is no longer a simple art form: *chartists* and *market technicians* are professionals who depend heavily on chart patterns and mathematical tools to predict buy and sell price points.

Chartists consider the different types of price charts as their primary tools and use charts exclusively as the basis of their market analysis. On the other hand, technicians do not solely rely on the use of charts. To outperform the market by forecasting the direction of asset price movements, market technicians use a combination of computerized test trading methods, selective market filtering and even obscure methods such as the study of astrology.

Oftentimes chart reading is based less on facts than is on intuition, thus different conclusions may be derived by different investors reading the same chart. What becomes the greatest challenge in the use of technical analysis is how well one makes sense out of price movement patterns and indicators. Finally, one frequently overlooked aspect of technical analysis is the role of risk management, especially in light of the 2008–2009 recession. Due to the subjective nature of chart reading, technical analysis poses a significant downside risk. Thus, regardless of the advances in trading tools and technology, any successful technical analyst has to combine chart reading and price forecasting with risk management strategies in order to mitigate risks and maximize gains.

Chapter 3

Chart Pattern Reading — Trend, Trend Line and Trend Channel

With our knowledge of the various styles of charts used in technical analysis from Chapter 2, how will a proficient technical analyst make use of the valuable information hidden in them? This chapter introduces the actual methodology of identifying chart patterns like trend, trend line and trend channel. Our investment decisions will depend on asset price forecasts by interpreting price and volume, as well as distinguishing different types of trends. We will look at different aspects of the charts in detail and make sense out of the columns and rows, lines and wiggles.

3.1 Trend

Spotting a trend in asset price movement is of the utmost importance in technical analysis. A trend is the continuation of price movement in a particular direction over a period of time, with the most generalized being upward, downward and sideways trends.

Asset price is often characterized by irregular cyclical movements, a series of zigzags that forms obvious peaks and troughs. A market trend is a pattern whereby asset price movements follow a generally upward or downward direction from present peaks or troughs over a defined, specific period of time. For example, an upward trend from a current trough may consist of several long upward trends with some short downward trends of shorter periods, and may represent an upward leg of an upward or downward trend over a longer period. Hence, it is imperative that one identifies different trends with reference to specific trend classification, otherwise "trends" spotted will yield little meaning or be misleading.

Technical analysts believe that asset price movements contain momentum and tend to move in trends. Once a trend is established, asset prices will likely follow in the same direction until it is met by an opposing force.

3.2 Trend Direction and Volume

Trend moves in three directions. An upward trend (see Figure 3.1) is defined as a series of peaks and troughs occurring higher in price successively through time; conversely, a downward trend (see Figure 3.2) is a series of peaks and troughs occurring lower. A sideways trend (see Figure 3.3) is a compromise between an upward trend and a downward trend — no significant price movements take place and a series of peaks and troughs follows a horizontal direction.

Trading volume may be used for trend identification to confirm a reversal pattern. In general, when the price of an asset is on an upward trend, its trading volume expands as price increases, and contracts when price decreases. Conversely, on a downward trend, trading volume and price changes have an inverse relationship. Volume relative to price movements can serve as a confirmation as to when a trend reversal is slated to occur.

Figure 3.1: Illustration of an upward trend

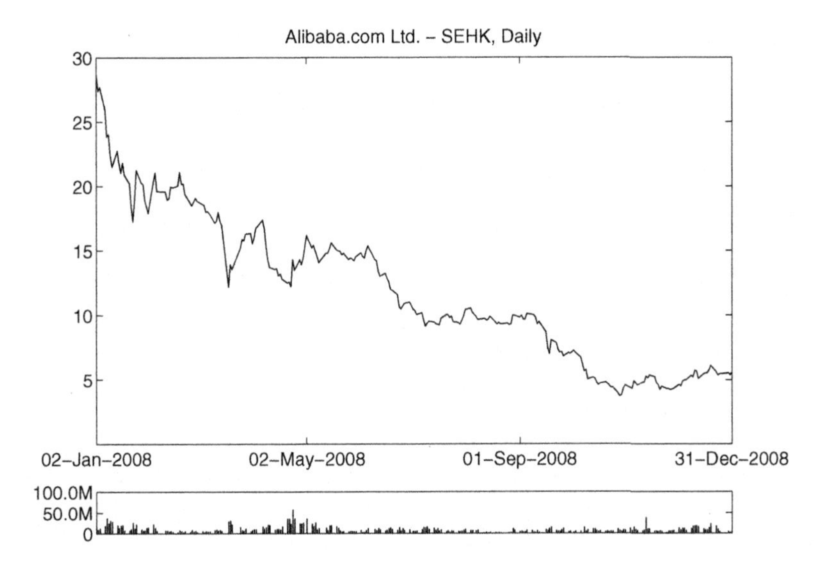

Figure 3.2: Illustration of a downward trend

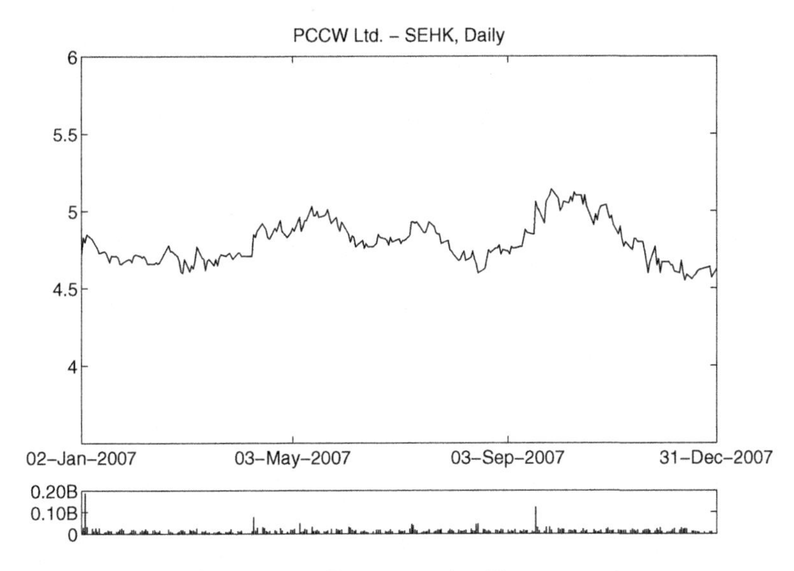

Figure 3.3: Illustration of a sideways trend

3.3 Trend Classification

Trends can be classified according to time-span into the following:

- Primary/Major trend: 1 to 3 years
- Secondary/Intermediate trend: 1 to 12 months
- Short-term trend: less than 1 month

The time-spans as shown above are only a rough guide, as there are no hard and fast rules to define the length of the time-span for a specific trend. More often than not, investors have the prerogative to set the length of time-span according to their objectives. This is especially true for short-term trends; the definition of a "short-term" time-span often depends on the patience and risk-tolerance preferences of individual investors.

How does trend classification assist investors in decision making? For example, if the long-term (primary) trend is up and the mid-term (secondary) trend is down, we should buy the asset now as its current price is low and has the potential to climb higher in the future.

Figure 3.4 shows a major downward trend for CITIC Pacific Ltd. over a 12-month period from January 2008 to December 2008. Even within this 12-month period, one can further divide the downward trend into several short-term upward and downward trends, with reference to different time segments. For example, a short-term downward trend occurred in

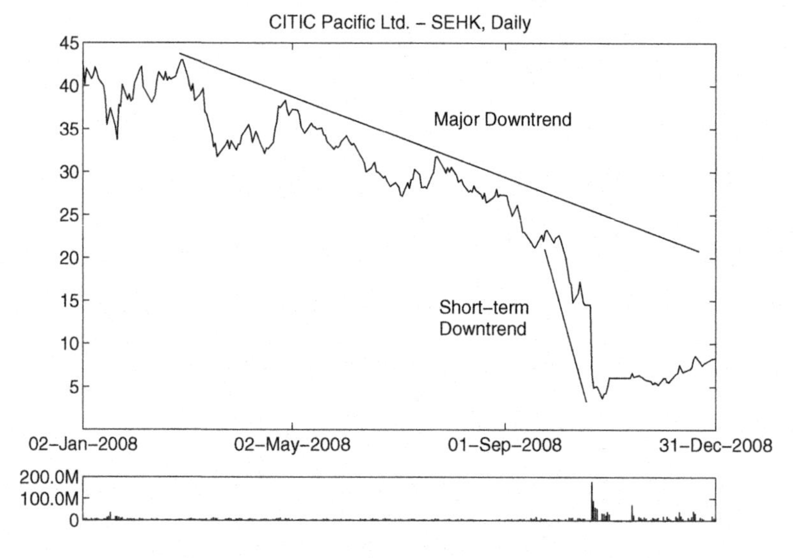

Figure 3.4: Illustration of trend classification

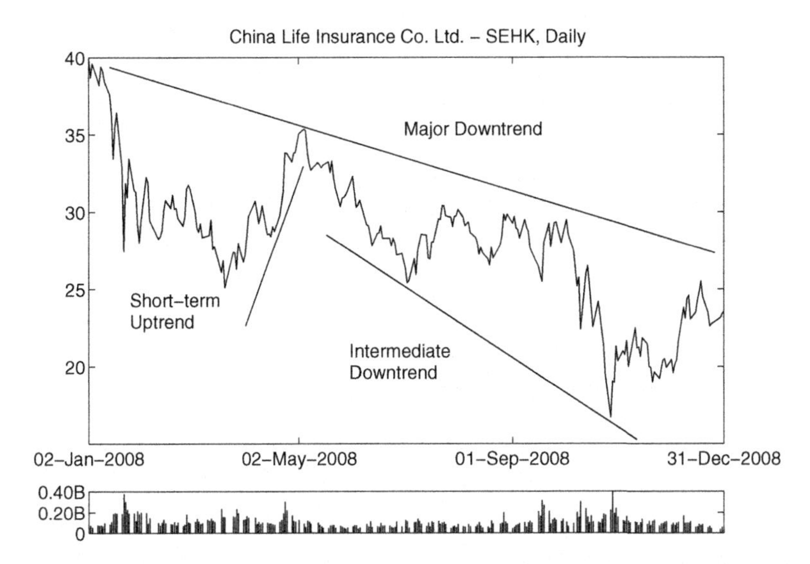

Figure 3.5: Another illustration of trend classification

September 2008 due to CITIC Pacific's investments in a currency accumulator linked to the Australian dollar. Another illustration of trend classification is given in Figure 3.5.

3.4 Support and Resistance

We understand that within a trend, a series of peaks and troughs is reached. Since price reversal takes place at these levels, they are defined as support levels and resistance levels, respectively. For instance, local minima and local maxima are often considered as support and resistance, respectively. From the above, one can therefore understand that the support level is a level drawn from previous lows over a period of time, and the resistance level from previous highs. In Figure 3.4, CITIC Pacific Ltd. dips to a all-time low \$3.66 owing to the accumulator incident. This price level is then a strong support.

To conceptualize our new terms, a trend is maintained by a series of support or resistant points sequentially heading in the same direction. For example, in a downward trend, the support and resistant levels show a descending pattern. In other words, if a correction causes the current asset price to rise higher than its previous resistance point, this may indicate that the downtrend is waning, and an impending uptrend may be on the horizon.

Once a support or resistance level is penetrated or broken through, its role is immediately reversed — the previous support level is now a resistance level for any future price movements, and vice versa.

Resistance and support levels are often maintained due to self-fulfilling trading methodologies. Investors are more willing to hold on to and buy assets when their prices fall below the established support level. Likewise, investors are keen to sell assets when asset prices advance to the established resistance level. Thus, resistance and support levels are a testament to the existing market belief over a period of time, upheld by "herd" mentality. One can take advantage of this information to trade one's assets effectively, especially in short-term profit taking.

Again, both support and resistance subject to a time period. Support and resistance can be further classified into short-term, mid-term, and long-term. This is analogous to the classification of trends. The importance of support and resistance levels relies on several factors. Firstly, the longer the period of time an asset price is traded at a support or resistance level, the more significant that level becomes, i.e., long-term levels are more important than short-term levels. Secondly, the more recent the support or resistance level is formed, the more likely the level will be conserved and unbroken. Lastly, the heavier the trading volume at the support or resistance level, the more vital the level is to investors, as trading activity prevents the level from being breached.

3.5 Objective Price Forecasting

Objective price is the minimum price level required for an upward or downward trend to be established. Usually, the asset price will advance or decline more than the objective price. However, the objective price is not used as a buying point or a shorting price. The objective price is not a target price, whereby it touches and retraces back. In trading, a trend is more useful than a price point. Instead of trading on the objective price, it is better to use the information this price provides to predict the conclusion of an uptrend or downtrend.

Interestingly, once the support or resistance levels are broken, one can expect an asset price change with magnitude equivalent to the difference between the support and resistance levels, e.g., with a support level at \$1.5 and a resistance level at \$2, should an uptrend penetrate the resistance level, one would anticipate asset prices to reach a minimum of $\$2 + (\$2 - \$1.5) = \$2.5$, as shown in Figure 3.6. However, this phenomenon may not apply to triangle pricing patterns (to be discussed in Chapter 4).

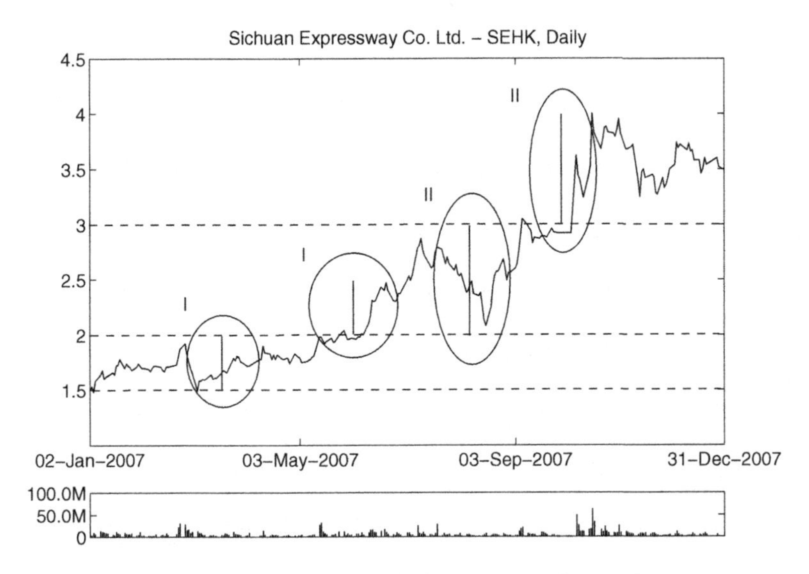

Figure 3.6: Examples of objective price forecasting

3.6 Round Numbers as Support and Resistance Levels

Sometimes asset price movements show strange characteristics, and price movements in stock markets are no exceptions. Equity markets are often marked by the "round number" phenomenon, a psychological support or resistance level based around important round numbers. For example, there appears to be a tendency for the whole Singaporean stock market to halt its advance or decline when the market index reaches certain points, particularly at multiples of 50. Understanding the significance of those points is immensely helpful in analyzing broader market trends, and is in turn vital for market timing.

Figure 3.7 illustrates Hang Seng Bank Limited with "round number" support and resistance levels in 2008.

3.7 Trend Lines

The trend line is considered as one of the most valuable tools in technical analysis for its ability to act as a time and price filter. It is a straight line that joins three or more successive peaks or troughs occurring over a period of time; see Figure 3.8 for an illustration. Although a straight line requires only the link between two separate points, a third point of contact

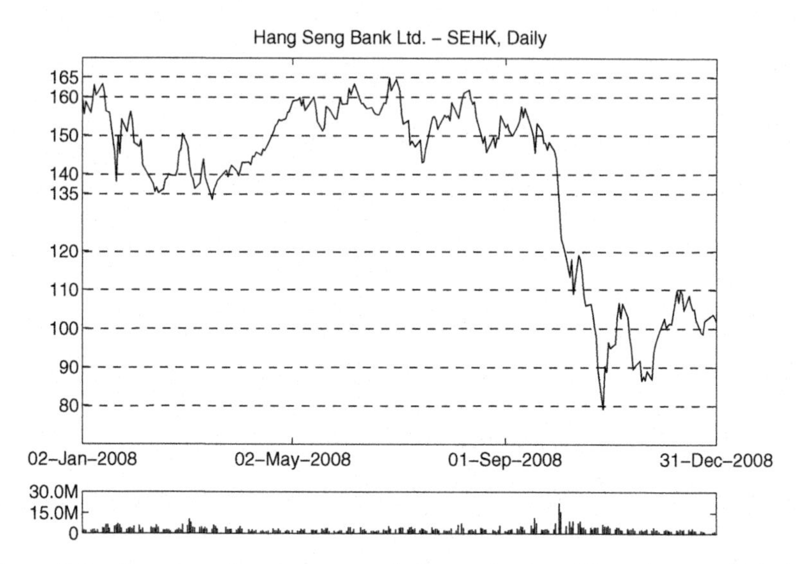

Figure 3.7: Round number phenomenon in the stock price of Hang Seng Bank Limited in 2008

Figure 3.8: An example of a trend line

is required to confirm a valid trend line. Obviously, the more points the trend line can be drawn through, the stronger the trend line — it is similar to resistance and support levels in this respect.

However, a prolonged trend indicates that the rate of price change has slowed down. Drawing from the concept of changes in velocity in physical objects, as acceleration approaches zero, an object's speed will remain constant. In similar fashion, a decrease in the rate of price change suggests a lack of force in the direction of the trend and may be a sign of reversal. The probability of a trend change is even higher once a trend line and a resistance or support level are broken in unison.

Once a trend line is verified as reliable and continues to move in the same direction, it allows technical analysts, traders and investors to surmise market sentiment at large. Trend line enables the prediction of impending asset price movements, thereby determining the timing for buy and sell signals.

3.8 Trend Channels

It is often the case that, with asset prices over a period of time, fluctuation occurs within the domain of a price ceiling and a price floor, such as the resistance and support levels. The same logic may be applied to trend lines. Once an initial trend line is drawn, one may observe that asset prices move in a range that is bounded by the initial trend line and another trend line that is almost parallel to it. For example, if a trend line is drawn on the connecting troughs of an uptrend, a parallel trend line may be drawn for the peaks of the same trend, as shown in Figure 3.9.

The range of price movement between the two trend lines is known as the trend channel. The trend line joining the peaks of a trend (upper line) is called the channel line or return line. It is called so because asset prices will stop advancing at this level and return to the lower limit of the channel. Essentially, the return line and the basic trend line are akin to resistance and support levels angled toward the direction of the prevailing trend. There are several forms of trend channels in technical analysis, including Raff regression channel, standard error channel and standard deviation channel.

Also, a price reversal can be anticipated by looking for channel breakouts. A fall below the support level indicates that the prices are likely to decline (see Figure 3.9), or an impending trend reversal. On the contrary, a breakthrough of the price movement above the resistance level signifies the forthcoming of a bullish market (see Figure 3.10). However, these breakouts tend to occur after certain warning signs; e.g., in an upward trend, one may

Figure 3.9: An example of a trend channel as support and resistance levels for price movement

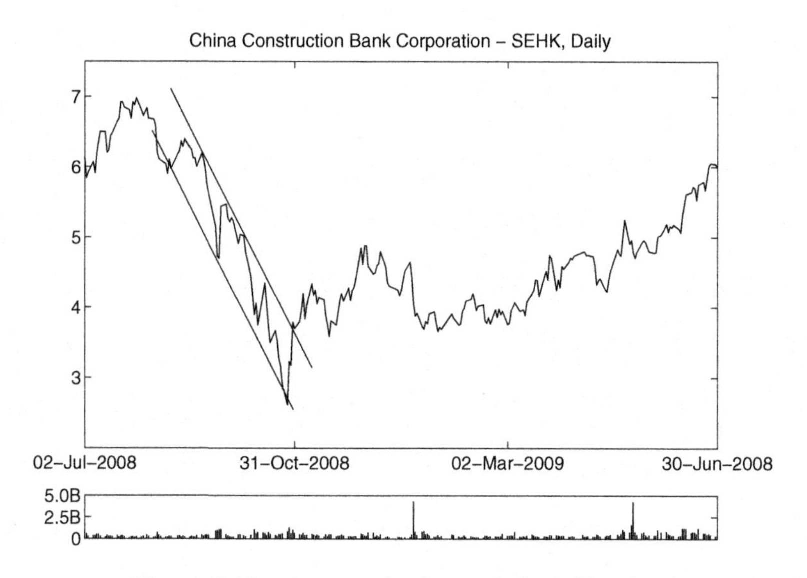

Figure 3.10: An example of a trend channel breakout

observe that the price movement has failed to reach the channel line, just shortly before a breakout actually takes place.

The ability to read the trend line and its trend channel enables the investor to trade his stock effectively. If he chooses a long position, he can buy his stock whenever the price falls to the support level, and sells it whenever the price advances to the resistance level. Or for those investors who are prepared to take a greater risk and choose to go for a short position, he can sell whenever the price touches the channel line, and buy whenever the price retreats to the basic trend line. This approach can be very useful for profit taking, especially so in short term stock trading.

Similar to trend channels, resistance and support levels can also be formed with moving averages. The moving averages will be discussed in Chapter 5, but simply put, they are the smoothened versions of the asset price movement. When using a moving average of a particular time interval, one can form a band by establishing levels above and below the moving average line with a set percentage, such that approximately 90% of the asset price movement is captured within the band. This is known to technical analysts as a price envelop. The envelop serves the same function as a trend channel, i.e., it acts as significant resistance and support levels. A special type of price envelop known as the Bollinger bands will be covered in Chapter 11.

3.9 Summary

The basic trend line is by far the most important and fundamental tool that has proven to be reliable in technical analysis. The channel line, although not as common and readily constructed on a price chart, also works pretty well in technical analysis. These tools therefore form the "must-have" in a serious investor's toolkits.

Also, certain advanced techniques and indicators in technical analysis are actually built on the concept of trend line or trend channel. Another example is the Gann's trend lines, which will be covered in Chapter 10.

Chapter 4

Chart Pattern Reading — Identifying Important Chart Patterns

In Chapter 3, we have discussed aspects of the asset price trend, and methods of identifying a price trend, a trend line, and a trend channel. We have also learned that the ability to gauge the asset price movement within a trend channel can be advantageous to investors and technical analysts in making their buying and selling judgments.

Although the study of price trends is useful even when market conditions are rather stable with a steady progressive price movement, we have in most part concluded that signs of price reversals, not price continuations, are our primary concern in analyzing trends. In other words, we should be concerned with fresh factors or adverse news that may cause a breakout from the existing trend. Such news could involve fundamental factors like the announcement of financial results, a merger or acquisition, economical factors like an impending interest rate hike, or the release of an inflation indicator figure.

Rarely do changes in trend occur suddenly. Often an important change in a prevailing trend undergoes a transition before it manifests into another distinctive trend. This intermediate step does not necessarily result in a trend reversal — it could also be a continuation of the prevailing trend but at a different pace and magnitude. The latter situation is called the consolidation phase.

Hence the study of price trend during its transitional period is yet another challenge for investors and technical analysts. It involves the identification of various chart patterns, analysis of their forecasting implications, and prediction of the magnitude of future price movements. In this chapter, we will focus on some important chart patterns that may develop during this phase.

4.1 Identifying Important Price Chart Patterns

Technical analysts believe that a chart pattern offers broad, revealing clues on asset price movement — but what exactly are price chart patterns and how do they differ from our knowledge of price trends? Essentially, a price chart pattern (less frequently known as chart formation) is a "picture" formed by asset price changes plotted on a chart over a period of time. A single pattern may involve any number of individual price trends. They can be grouped into two distinct categories: those that signal a trend reversal and those that suggest a trend continuation, appropriately named reversal pattern and continuation pattern, respectively.

4.2 Major Reversal Patterns

The followings are the major reversal patterns that all investors must know.

Shoulder–Head–Shoulder (SHS) Pattern

This is one of the most famous reliable reversal patterns. This pattern is characterized by its three well-defined peaks distributed along three different time segments. The central peak is the highest of the three, while the two peaks on both sides are typically at a similar price level to each other. As the price chart pattern resembles the outline of the head and shoulders of a human being, it is referred to as the "shoulder–head–shoulder" (SHS) pattern, or simply the "head and shoulders" (H–S) pattern. An example of an SHS pattern is shown in Figure 4.1.

An SHS pattern begins with a sharp price rally before dipping back to form the left shoulder. The pullback is not unusual and may be the result of a number of factors including profit taking or panic selling as a result of adverse market information.

After a discernible amount of time and the asset price retracts to its support level, investors who have missed the first rally take hold of the fact that the price has fallen to an affordable level, and they thus "seize the opportunity" or embark on their "bargain hunting" in the market. This causes the price to advance again, but this time reaching a higher level than the previous high. Then there could be bad news again, either due to fundamental factors or economic results, making the price drop below the

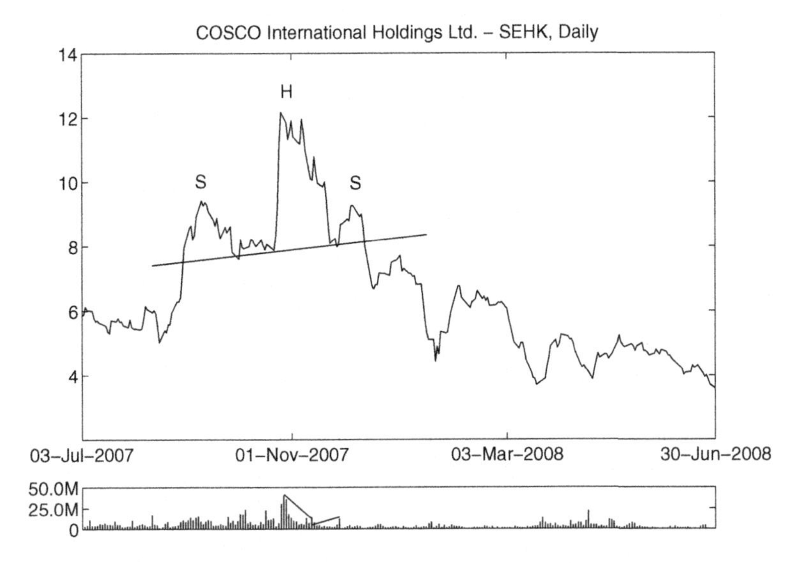

Figure 4.1: An example of an SHS pattern

previous peak attained by the left shoulder. This move realizes the "head" of SHS pattern.

Again, the tardy investors take the opportunity to enter the market and so the price rebounds but fails to reach the previous peak. Soon after, the price slips and thus completes the right shoulder of the SHS formation.

While observing and recording the occurrence of an SHS pattern, one should note that the most distinguishable feature is the neckline. It is a near horizontal line drawn between the low upon which the left shoulder is formed, and the low prior to the formation of the right shoulder. The neckline serves as a support level of the pattern, and may be slightly tilted upwards or downwards depending on the prior trend.

Once the asset price falls through this neckline, it signifies a potential trend reversal. One may also notice that after the asset price penetrates the neckline, it tends to rebound back toward the neckline, but fails to breach that level and will continue to move lower as the neckline takes on the role of a resistance level. An SHS pattern can only be considered "complete" when the neckline is decisively broken through by about three percentage points in terms of price movement.

In another interesting phenomenon, the minimum price fall below the neckline may be estimated by measuring the vertical distance from the top of the highest peak (the head) to the neckline. This vertical distance is then projected downwards from the level where the neckline is broken, as

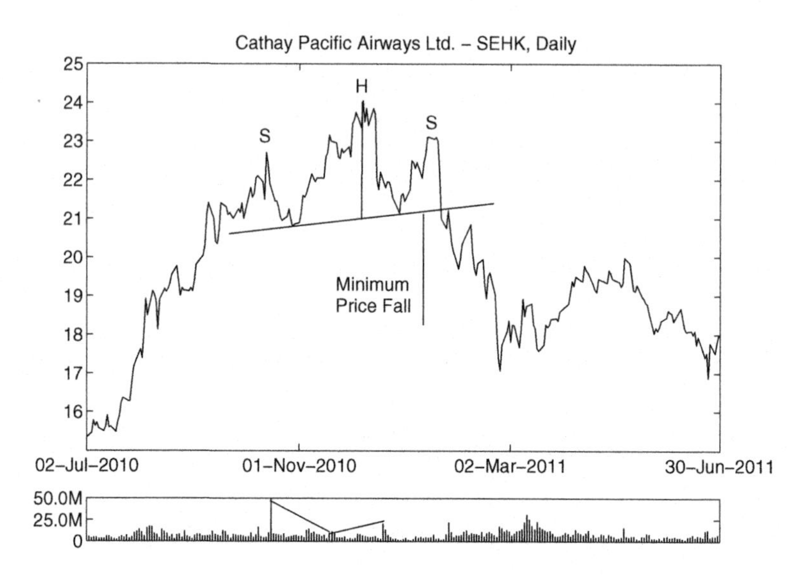

Figure 4.2: Estimating the minimum price fall with an SHS pattern

illustrated in Figure 4.2. Technical analysts often find this approach helpful in determining the extent of a price drop whether or not to enter a position. A horizontal neckline tends to give a "more reliable" minimum price fall approximation than the tilted ones, since it is easier to measure accurately the vertical distance between the highest peak and a horizontal neckline rather than the mathematical average level of a tilted one.

In addition, the volume traded at the different phases of an SHS pattern is rather important in the study of its implication. In general, the volume is higher as the left shoulder develops, slightly less as the head is formed; and low as the right shoulder takes shape. This is largely due to the "drying up" of buying power, which ultimately leads to the decline in asset prices.

The formation of an SHS pattern varies depending on the volatility of the asset, spanning from as short as one month to as long as a few years. In general, the longer the time period taken to form an SHS pattern, the longer the resulting bear trend is likely to be.

Inverse SHS Pattern

This pattern is the mirror image of the SHS pattern flipped along the horizontal axis. It has three distinct bottoms, i.e., the head formed by the middle trough, and a shoulder each on its left and right with slightly higher troughs, as depicted in Figure 4.3.

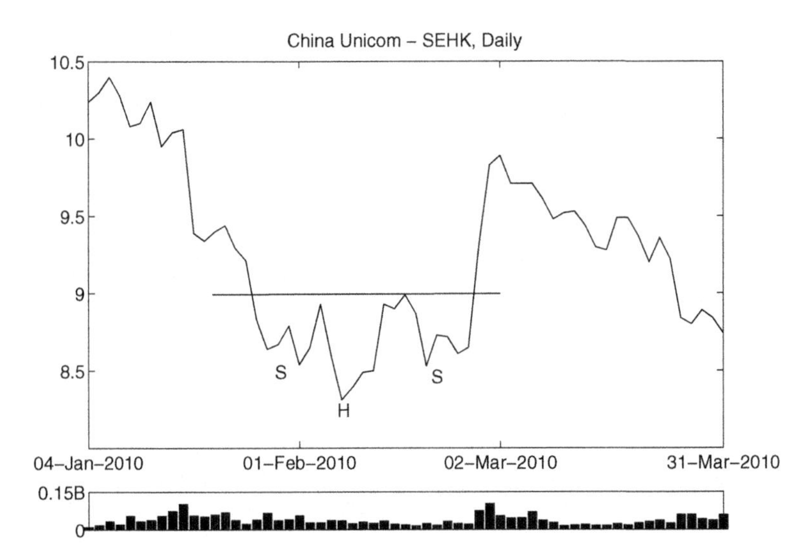

Figure 4.3: An inverse SHS pattern formed in early 2010 by China Unicom listed on SEHK

Although similar to the SHS pattern in many respects, the inverse SHS pattern can only be considered "complete" after a decisive breach of the neckline, with an additional condition that trading volume dramatically increases towards the end of the right shoulder.

The trading volume during the two phases of the inverse SHS pattern is usually light on the left shoulder, even lighter at the head until the trough of the head is reached, then an increase for the return to the right shoulder, whereby volume becomes very light again. The critical point occurs after the right shoulder trough when the price rallies with a sharp burst in trading volume. In addition, the pull back toward the neckline after the breakthrough, also called the return move, is more likely to take place compared to the normal SHS pattern.

Complex SHS Pattern

Typical SHS and inverse SHS patterns consist of a single left shoulder, a single head and a single right shoulder. However, variations on the SHS pattern may involve different number of shoulders or steep neckline slope. See Figure 4.4 for illustration of a complex SHS pattern.

Technical analysts may note that there is a strong tendency for symmetry in the formation of the SHS pattern. If a single left shoulder occurs, it is highly likely that a single right shoulder will develop later. In the case

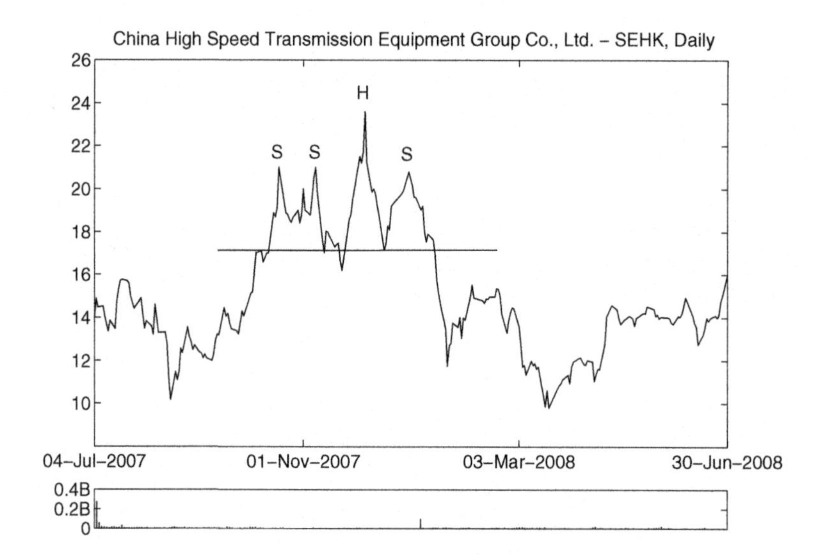

Figure 4.4: An example of a complex SHS pattern

of double or triple left shoulders, the probability of double or triple right shoulders appearing would be higher (but not necessarily).

As mentioned earlier, the more horizontal the neckline, the more predictable the price movement will be exiting the SHS pattern on the reverse side of the neckline. In other words, for an SHS pattern with a neckline tilting upward at a large angle, the subsequent reversed trend will be less significant.

Double Top and Double Bottom Pattern

Double top and double bottom pattern are also distinctive sets of reversal chart patterns. The pattern can be easily recognized with its unique twin peaks or bottoms resting on top of a neckline and separated by a deep valley. The peaks or bottoms usually reach the same price level at the reversal point. Thus, the double top pattern takes the shape of the letter "M", and the double bottom pattern resembles the letter "W". An example of a double top pattern can be seen in Figure 4.5.

In terms of trading volume, activity tends to be heavy during the first peak of the double top pattern, and lighter on the second peak, and vice versa for the double bottom pattern, i.e., volume is light on the first trough and heavier on the second trough. Once again, it is important to observe increasing momentum on asset prices upon the breakthrough of the neckline before we can confirm that the pattern has been completed.

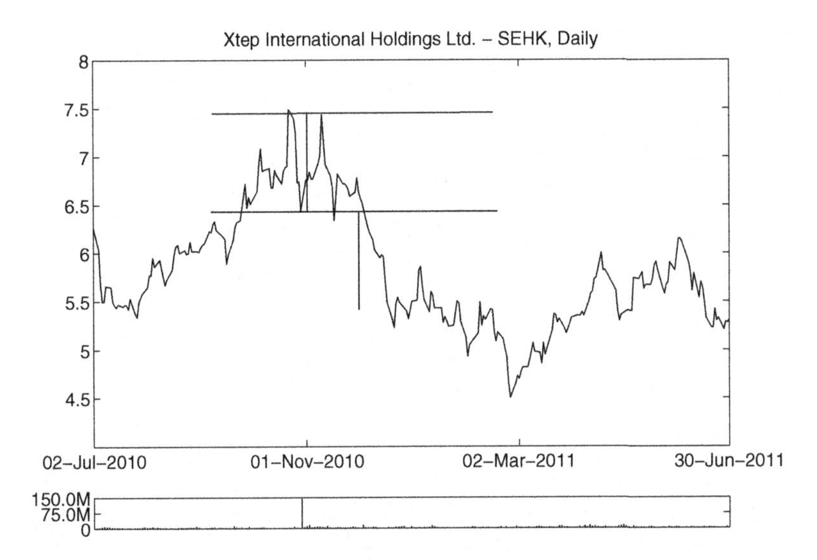

Figure 4.5: A double top pattern

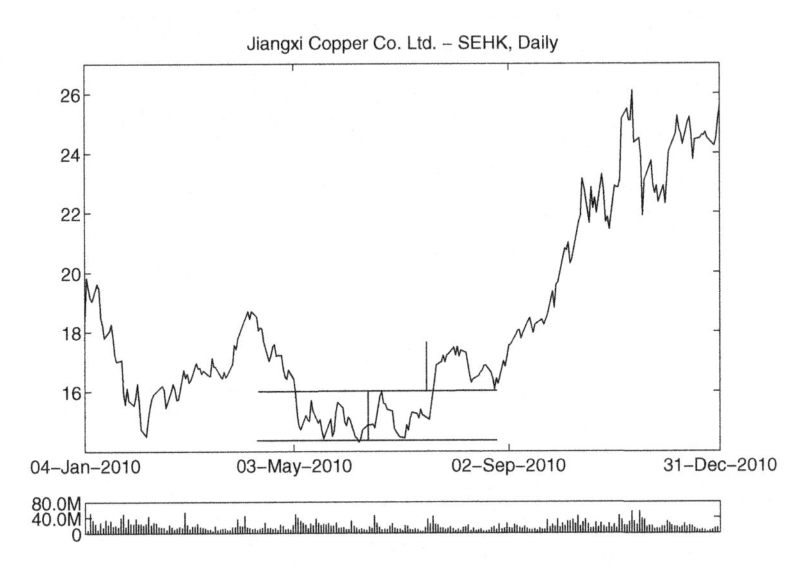

Figure 4.6: A triple bottom pattern

Some double top and double bottom patterns could even extend to form multiple top or bottom patterns, as illustrated in Figure 4.6. Nevertheless, we may apply to double top and double bottom patterns, the same approach used in SHS patterns to predict the magnitude of the reverse trend,

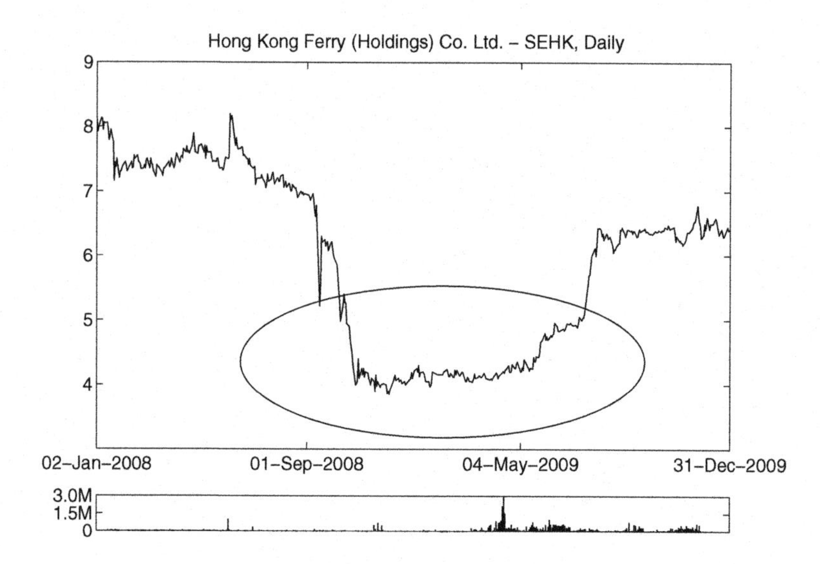

Figure 4.7: A saucer pattern formed between October 2008 and June 2009

i.e., vertical projection from the head to the neckline and from the neckline to estimated endpoint.

Rounding Bottom or Saucer Pattern

This pattern does not occur as frequently as the patterns mentioned above. It is formed by a gradually changing trend. At market bottoms, this pattern resembles a saucer or a bowl as depicted in Figure 4.7, and at market tops, it is usually referred to as the "inverted" saucer.

The saucer pattern does not provide a clear identification of support or resistance levels, and thus there exists no definite methods to predict impending price movement in the reverse direction. It is also not easy to identify at which point the saucer pattern has been completed, especially for patterns that span years. As seen from Figure 4.7, the saucer pattern developed over a time frame of 9 months, from October 2008 to June 2009. One conclusion is that the longer it takes to complete the pattern, the greater the potential for bigger price movements in the reverse direction.

4.3 Major Continuation Patterns

Continuation patterns represent pauses in the current trend, not necessarily indicating that a price reversal is in progress. In this section, we will cover some of these patterns, namely the rectangle pattern, triangle pattern, flag

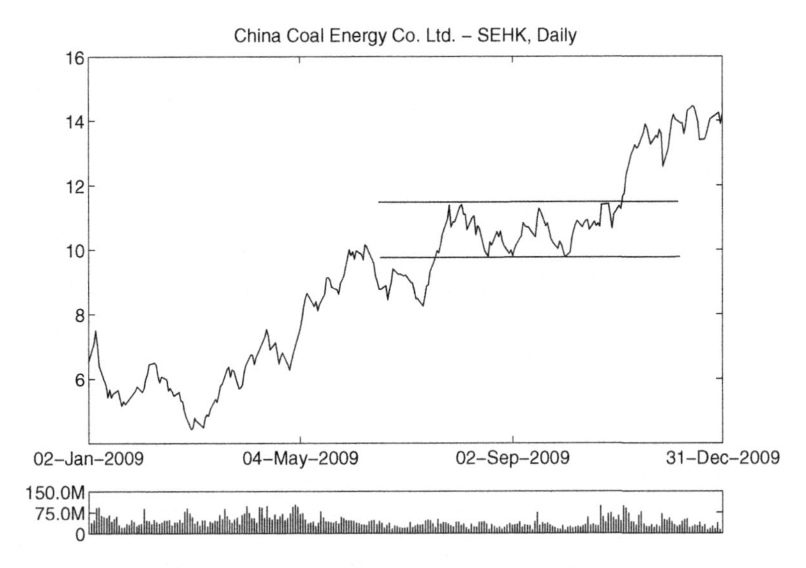

Figure 4.8: An example of a rectangle pattern

pattern, pennant pattern, and wedge pattern. In addition, a brief mention will be made on the "gap" pattern.

Rectangle Pattern

This is the box-like pattern that emerges when the prices swing repeatedly between certain support and resistance levels. See Figure 4.8 for an example of a rectangle pattern. The formation of this pattern represents a consolidation phase before resumption of the current trend, or a stepping stone to a longer-term development.

Price movement upwards through the rectangle's resistance line, if significant, would indicate a bullish trend. Alternatively, any decline through the support line would signal further weakness. These significant breakouts indicate the completion of the rectangle pattern, and points to the direction of the coming trend; the resulting price fluctuation can be expected to occur quickly and have a magnitude of at least the same width of the rectangle.

Investors therefore can take advantage of price swings by buying when the price dips near the bottom of the range, and selling when it rallies near the top, similar to trading within a trend channel. This approach, which is sometimes called the "counter-trend approach", is good for short-term trading, especially so since the range is well defined, and the risk is relatively less, so long as the trading range remains intact.

However, an investor must keep a close watch when he perceives that a rectangle pattern is developing, as the rectangle pattern could well turn out to be a multiple top or bottom pattern, thus signaling a possible trend reversal.

Triangle Pattern

In a triangle pattern, two converging lines are drawn at the top and bottom of the price movement "coil" when the price movement over time is one of decreasing fluctuation, such that the apex of the triangle pattern is always to the right of its base. In other words, we have an isosceles triangle lying on its side. In most cases, triangle patterns are continuation patterns, although they sometimes turn out to be reversal patterns, depending on whether the subsequent breakout is through the top or the bottom edge of the triangle, with respect to the prevailing trend prior to the development of the triangle pattern.

There are generally three types of triangle patterns.

(1) Ascending Triangle. The ascending triangle has a rising lower side, and a horizontal or somewhat flat upper side. An ascending triangle is generally a bullish pattern.

(2) Symmetrical Triangle. A symmetrical triangle has two converging trend lines, with the upper line descending and the lower line ascending. It is also know as the normal "coil".

While a symmetrical triangle is being formed, the trend is expected to continue; and a breakout through either the upper line or the lower line is invariably at the point after $\frac{1}{2}$ or $\frac{3}{4}$ of the triangle has already taken shape, i.e., before the triangle is fully formed.

(3) Descending Triangle. This has the upper side descending, and the lower side horizontal or almost flat. A descending triangle is usually a bearish pattern.

Figure 4.9 depicts the ascending, symmetric triangle and descending triangle patterns. A triangle pattern is considered complete upon a decisive breakout of its upper or lower sides.

Flag Pattern and Pennant Pattern

A flag pattern looks rather similar to the rectangle pattern, both having the upper and lower trend lines and appearing in the shape of a parallelogram. The difference is the direction of the trend lines forming the

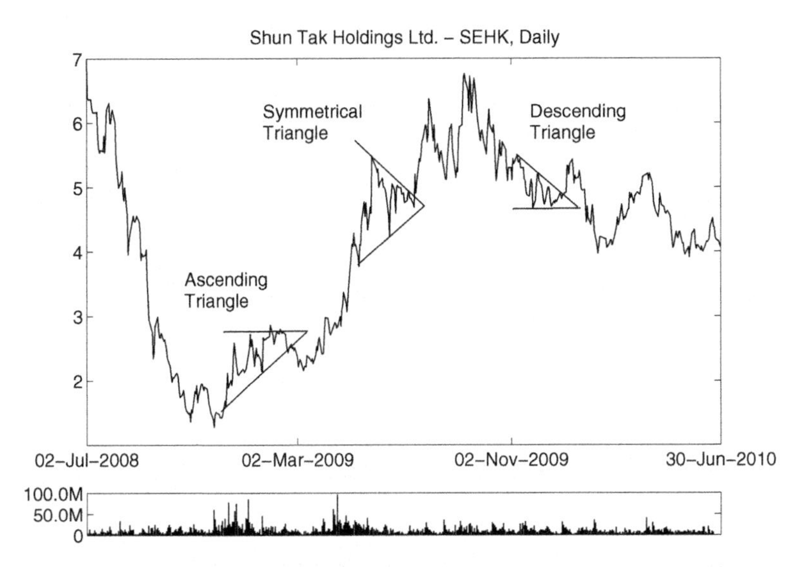

Figure 4.9: Examples of the ascending, symmetrical and descending triangle patterns

parallelogram; whereas the rectangle pattern has its trend lines almost horizontal in alignment, the flag pattern has its parallel trend lines sloping upwards or downwards from the horizontal line.

A flag pattern occurs as a short pause in a dynamic market where there is a rapid advance or decline in stock prices. Its occurrence may be referred to as a situation in which "the market is trying to catch its breath", before advancing or declining further. The rapid advance or decline as plotted on the price chart is often steep, and the steep price movement thus forms the "flag pole" for the flag pattern.

A flag pattern is characterized by having its slope against the prevailing trend direction. Thus, in an uptrend market the flag is sloping downwards, whereas in a downtrend market, the flag is sloping upwards. Also, the trading volume tends to be thinning during the period the flag pattern takes shape. And if an upward price movement occurs, its range of fluctuation is expected to be as much as that of the width of flag. This price movement is all the more significant if accompanied by heavy volume.

In a downtrend market, the flag pattern, which is upward bias, tends to be accompanied by a declining volume even though the price can be rising during the flag formation. And once the price falls through the flag, the price slide would continue at almost the same pace and amount as before its formation. Volume tends to be higher during the price fall, but may not be as rapid as in the case of an uptrend market.

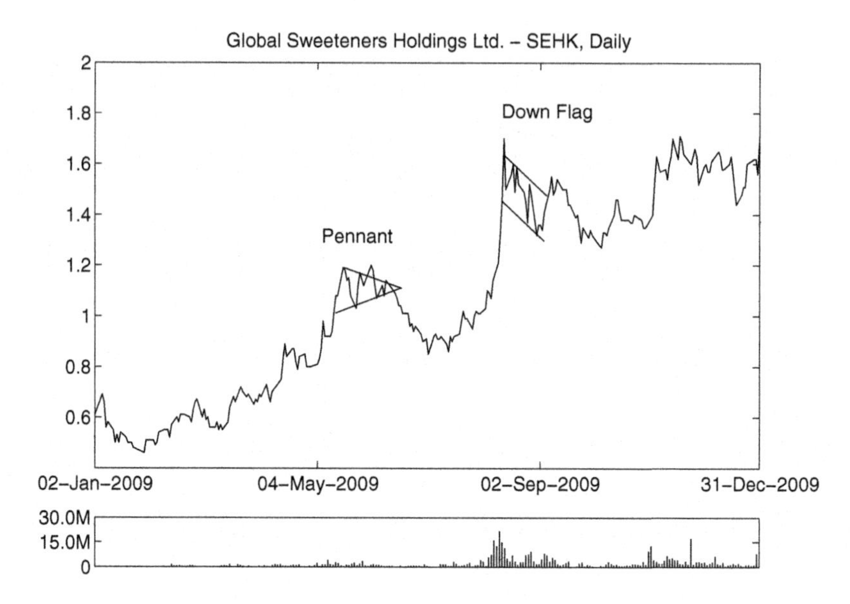

Figure 4.10: Examples of a down flag pattern and a pennant pattern

A pennant pattern resembles a symmetrical triangle but smaller in size and shorter in duration. This pattern has similar characteristics as the flag pattern in terms of trading volume, time frame for it to take shape, and the subsequent price movement upon breakout of the pattern.

Illustration of the flag and pennant patterns can be found in Figures 4.10 and 4.11. The flag pattern or the pennant pattern usually develops within 1 to 4 weeks. They are often seen as reliable and useful for short term trading.

Wedge Pattern

A wedge pattern appears somewhat similar to a triangle pattern. It also consists of two converging lines drawn through the peaks and troughs during a specific period. The close resemblance at times leads to confusion among investors. The triangle pattern has three combinations depending on their converging sides, and the direction of the apex. On the other hand, the two major wedge patterns have

(1) two rising lines with different gradients, or
(2) two falling lines with different gradients.

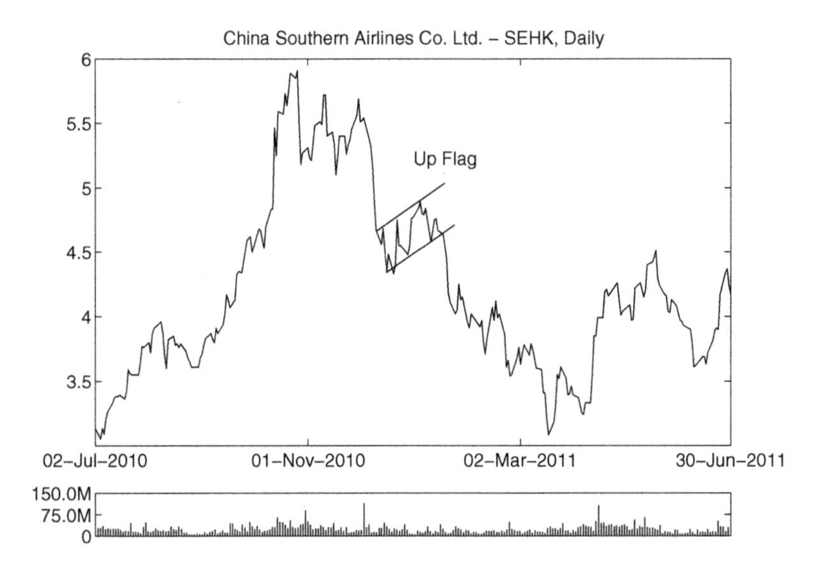

Figure 4.11: An example of an up flag pattern

And its apex, although pointing toward the right, it is not horizontally. A comparison between triangle patterns and wedge patterns is illustrated in Figure 4.12.

A wedge pattern represents a pause or a temporary consolidation in the current trend. It can be formed within 2 weeks to 3 months. It is more noticeable in the daily or weekly chart, but not in the monthly chart.

Unlike the common connotation for the ascending triangle and descending triangle which signals bullish and bearish price movements respectively, the rising and falling wedges signal the opposite. In other words, a rising wedge is a bearish pattern and a falling wedge is a bullish pattern. This is because the wedge pattern also tends to slant against the prevailing trend, just like the flag pattern.

Hence, in an uptrend market, there is a tendency that a falling wedge pattern would occur, and when the price breaks through the upside, it would continue its upward movement, as can be seen from Figure 4.13. And usually, after the breakthrough, the price would fall back to find support at the upper side of the wedge. In a downtrend market, a rising wedge as illustrated in Figure 4.14 could be formed, and the price falling through the lower side of the wedge is likely to be fast, in the direction of the prevailing downward trend.

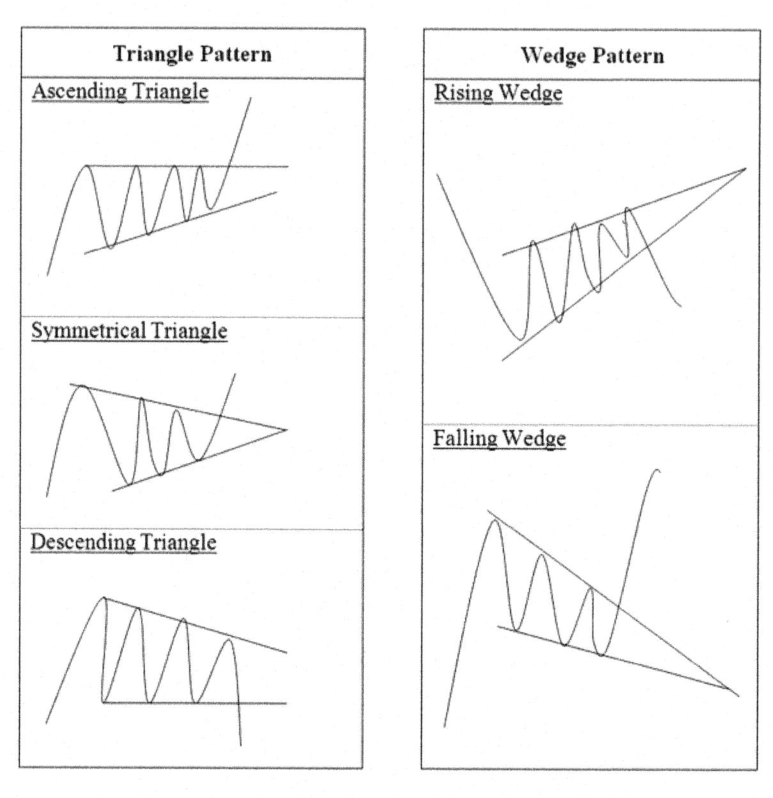

Figure 4.12: Triangle pattern and wedge pattern at a glance

Gap

A gap is formed when the trading range of a price bar for a day does not overlap with that of the previous day. In short, there exist a price range with no trading taking place. See the illustration of gaps in Figure 4.15.

A gap often happens in a rapidly rising market, when the stock price of a certain day opens above the highest price of the previous day; and the price chart so plotted will have a gap between the two consecutive trading days. Similarly, in a rapidly falling market, a gap occurs when the day's opening price is lower than the previous day's low.

A gap is considered "filled" or "closed" when the price later comes back to the price range. This process of filling the gap can take place within a few days, or as long as a few months. Quite often, investors believes that "a gap has to be filled", and that the occurrence of a gap would offer them another opportunity to enter the market, or to sell at a better price; although such gap filling may not necessarily happen.

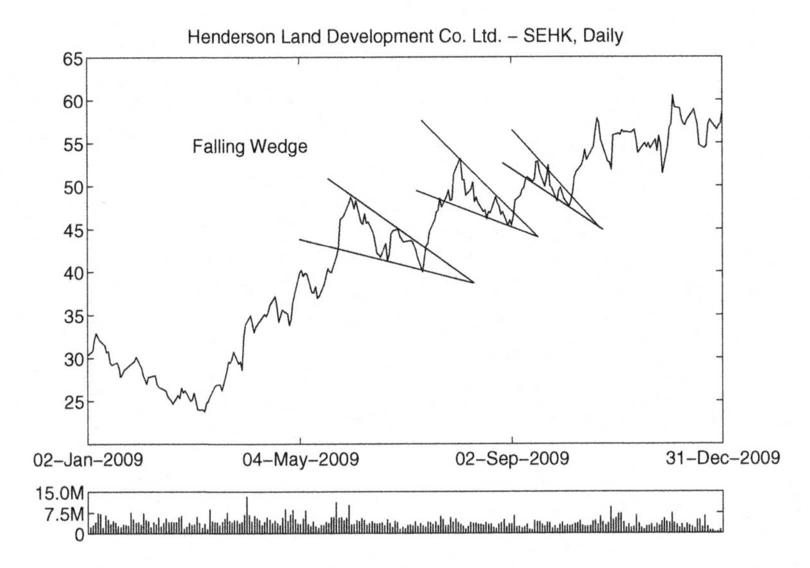

Figure 4.13: An example of a falling wedge pattern

Figure 4.14: An example of a rising wedge pattern

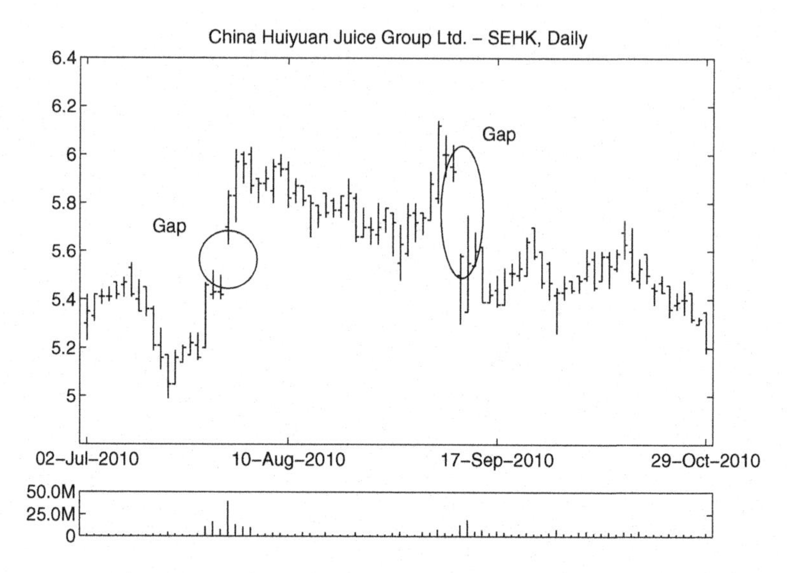

Figure 4.15: An example of gaps in the price movement

4.4 Summary

We have covered a total of 10 common chart patterns that are important and useful to the general investors. However, they are certainly not the be-all or end-all of chart patterns that can be found in the numerous price charts in the broad field of technical analysis. There are many more, but they are not so commonly used, and we would leave them to the readers to discover on their own [Pring (2002); Bulkowski (2005)].

The skill of chart pattern reading comes with experience, and it is often regarded as an art in technical analysis; since the interpretation cannot be quantified mathematically. Thus, the reading of chart patterns can be very subjective, and such subjectivity often leads to the saying that "the beauty of chart patterns lies in the eyes of the beholder". This is because, from a chart, one tends to see what one wants to see, or choose not to see what one does not want to see. Also, different persons reading the same chart could end up seeing different chart patterns altogether.

With the knowledge of technical analysis becoming more widespread, more investors are now familiar with the implication of the various patterns, and therefore will act according to what they see in the chart patterns. Besides, the availability of sophisticated computers and softwares have granted investors an easy access to information. This could lead to a concerted effort by a big pool of investors taking the same position at the same time. This could result in the so-called "self-fulfilling prophecy".

And it is the inherent subjectivity of chart pattern reading that leads us to examine the various stock market indicators in Part 2. Indicators are more objective and not as open to interpretation as chart patterns alone; as they are statistical constructs or models that aim at analyzing the stock market in more scientific ways, each with quantifiable input and output using mathematical and statistical variables. Some of these indicators will be covered in detail in the forthcoming chapters of this book.

PART 2

Filtering Tools in Technical Analysis

Chapter 5

Linear Filters

In the previous chapters, we have considered the chart pattern reading in technical analysis, such as trend lines, trend channels and various chart patterns. However, these kinds of figurative techniques can be subjectively interpreted by different traders. Therefore, practitioners came to an agreement that a universal and quantifiable trading tool should be developed.

The family of moving averages is one remarkable example of such quantifiable tools. Nowadays, different kinds of moving averages are actively used in the market by technical analysts. Visually speaking, moving averages resemble the smoothed versions of the original data. Trading strategies based on these moving averages are then designed, for instance, one can detect buying or selling signals by checking the crossover of two moving averages. Before understanding how this mechanism works, we first have to grasp the idea of linear filters.

5.1 Linear Filters

Filters are a common term in the mathematics and engineering fields. They transform input signals into output signals that fulfill specific purposes. In financial time series, moving averages are the output of the original financial data via linear filters. Linear filters can be mathematically quantified so that it is possible to study their properties, like why the output data is smoothed or shifted in time. In this chapter, we start from the introduction of linear filters. Linear filter is an important tool in discrete-time signal processing. A classic reference text on digital signal processing is Oppenheim and Schafer (1989). Books like [Ingle and Proakis (2000); Stearns and Hush (2011)] are a good supplement to the area of digital signal processing with many MATLAB examples. Moreover, linear filters in the form of technical indicators are discussed in technical analysis books like Ehlers (2001); Lee and Tryde (2012); Mak (2003, 2006); Meyers (2011); Pring (2002).

In this context, our input signal is simply the financial data. Financial data observed in real life are discrete-time signals, i.e., they can be indexed by a sequence of numbers. Throughout this book, we let the symbol $\{x_j\}$ denote a discrete-time signal, which is a function of integers j. The integer j represents time and runs from negative infinity to infinity:

$$\mathbf{x} = \{x_j\}_{j=-\infty}^{\infty} = (\ldots, x_{-2}, x_{-1}, x_0, x_1, x_2, \ldots).$$

In digital signal processing, filters turn the original input signal $\{x_j\}$ into another signal of interest, denoted by $\{y_j\}$ in this context. In particular, a linear filter is described as the convolution sum of the following form:

$$y_j = \mathbf{h} * \mathbf{x} = \sum_{k=-\infty}^{\infty} h_k x_{j-k},$$

where

$$\mathbf{h} = \{h_k\}_{k=-\infty}^{\infty} = (\ldots, h_{-2}, h_{-1}, h_0, h_1, h_2, \ldots)$$

is the processing tool known as filter. The filter is comprised of real numbers h_k's, also known as the weight, the filter coefficients or the impulse response. Note that y_j is a real number indexed by j. If we line up all the y_j's for all j's, we obtain the desired output signal processed by the filter $\mathbf{h}$:

$$\mathbf{y} = \{y_j\}_{j=-\infty}^{\infty} = (\ldots, y_{-2}, y_{-1}, y_0, y_1, y_2, \ldots).$$

We demonstrate the effect of a basic linear filter known as the centered moving average (CMA). It has the following form

$$y_j = \frac{1}{2M+1}(x_{j+M} + \ldots + x_{j+1} + x_j + x_{j-1} + \cdots + x_{j-M}),$$

where the filter coefficients h_k's are $1/(2M+1)$ for $k = -M, \ldots, M$ and zero otherwise. Note from here that strictly speaking, the CMA should mean the filter $\mathbf{h}$ with the mentioned h_k's. However, it is also customary to call the output signal $\mathbf{y}$ as the CMA. In the remaining context, we will stick to this fashion and refer to the output signal with the name of the filter. For instance, the notation CMA_{10} represents a CMA line with $M = 10$.

In Figure 5.1, we apply the CMA with $M = 10$ on the Hang Seng Index in 2011. The first and last 10 values are not visible because data beyond 2011 are not used in this example. From Figure 5.1, we observe that the CMA is the smoothed version of the original data.

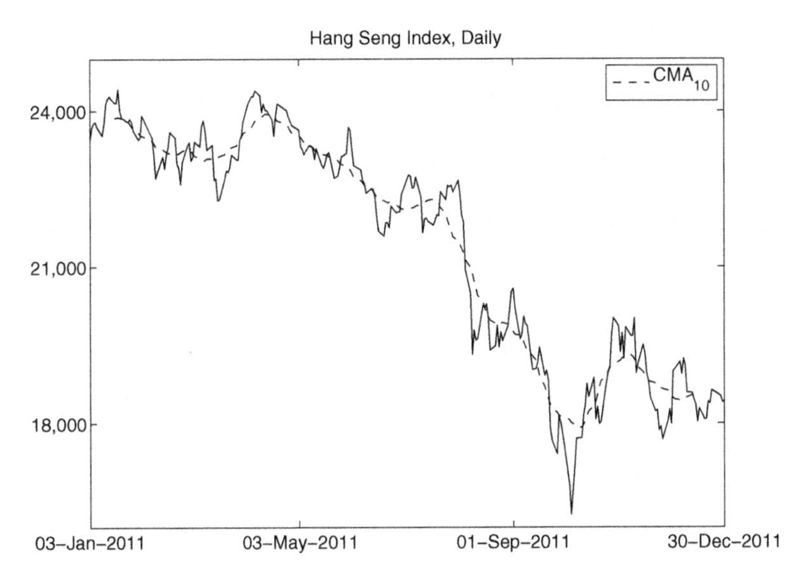

Figure 5.1: CMA with $M = 10$ applied on the Hang Seng Index in 2011

In reality, we are confined to the past data since no future data are available and all we observe is a financial time series up to a certain day j:

$$\mathbf{x}_j = (\ldots, x_{j-1}, x_j).$$

In such case, a linear filter is reduced to a causal form starting from $k = 0$. It gives an output signal

$$y_j = \sum_{k=0}^{\infty} h_k x_{j-k}.$$

The simple moving average (SMA) is the most fundamental causal filter. An M-period SMA is defined as

$$y_j = \frac{1}{M}(x_j + x_{j-1} + \cdots + x_{j-M+1}). \tag{5.1}$$

In Figure 5.2, we see the effect of a SMA with $M = 20$ acting on the Hang Seng Index in 2011. As seen from the figure, the beginning 19 values are missing this time because the index values before 2011 are not used. The SMA is a smoothed version of the original data, similar to the case of CMA. However, we observe this time an interesting phenomenon that the SMA is a little bit delayed in time. For the CMA, it is simply smoothed but not shifted in time. The reasons for these behaviors will be explained in Section 5.2 when we consider linear filters in the frequency domain.

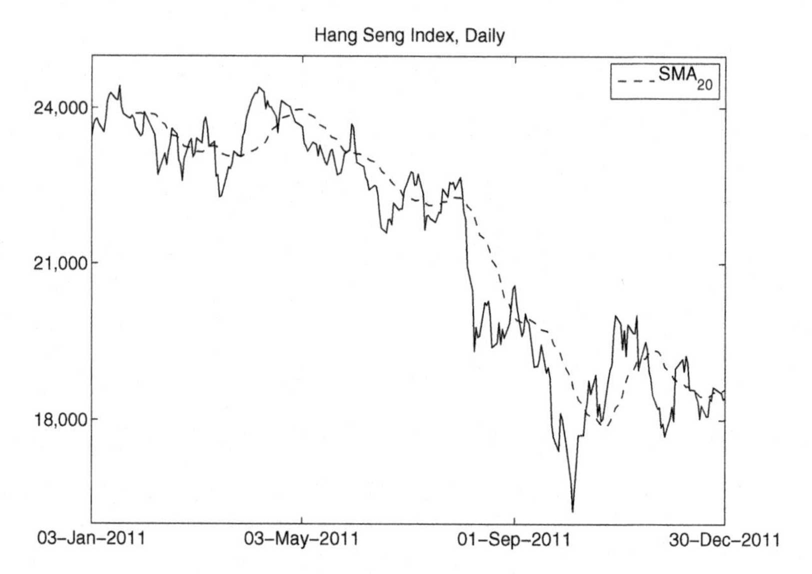

Figure 5.2: SMA with $M = 20$ applied on the Hang Seng Index in 2011

5.2 Frequency Response

In the last section, we have considered the filter behaviors in the time domain. To characterize a linear filter, we can also take the perspectives from the frequency domain. This is accomplished by studying the frequency response of a linear filter.

One way of understanding the frequency response is to use an analytic eigenfunction $\mathbf{x} = \{x_j\}_{j=-\infty}^{\infty}$, $x_j = \exp(ij\omega)$ with any frequency ω as an input signal. Note here that i is the imaginary unit and that is why we use j to represent an integer. Then, a linear filter will return the following output signal

$$
\begin{aligned}
y_j &= \sum_{k=-\infty}^{\infty} h_k x_{j-k} = \sum_{k=-\infty}^{\infty} h_k \exp(i(j-k)\omega) \\
&\equiv H(\omega)\exp(ij\omega) = H(\omega)x_j,
\end{aligned}
$$

where

$$
H(\omega) = \sum_{k=-\infty}^{\infty} h_k \exp(-ik\omega), \tag{5.2}
$$

is called the frequency response of the underlying system with regard to the filter $\mathbf{h}$. From (5.2), we see that the frequency response is given by the discrete-time Fourier transform (DTFT) of the filter coefficients h_k. It is well known that the Fourier transform gives the frequency domain representation of the time-varying input. Also (5.2) tells us that the frequency response $H(\omega)$ acts as the eigenvalue of the analytic eigenfunction, and eigenvalues are notable for stretching or shrinking the input. In linear filtering problems, the frequency response reveals the frequency characteristics.

Suppose we let $H_R(\omega)$ and $H_I(\omega)$ be the real and imaginary parts of the frequency response $H(\omega)$. We then write $H(\omega)$ into the polar coordinate system

$$H(\omega) = H_R(\omega) + iH_I(\omega) = |H(\omega)| \exp(i\phi(\omega)),$$

where

$$|H(\omega)| = \sqrt{H_R^2(\omega) + H_I^2(\omega)},$$

is called the amplitude of the frequency response and also known as the gain function, and

$$\phi(\omega) = \tan^{-1}\left(\frac{H_I(\omega)}{H_R(\omega)}\right)$$

is called the phase of the frequency response.

We can learn about the properties of a linear filter by studying its frequency response, or specifically its amplitude and phase function. Assume now we have an input signal $x_j = \exp(ij\omega_0)$ with a specific frequency value ω_0, the output signal is

$$y_j = H(\omega_0)x_j = |H(\omega_0)| \exp(i(j\omega_0 + \phi(\omega_0))).$$

The output signal is determined by the frequency response in the way that it is being amplified by the amplitude and phase-shifted from the original data by the phase function. For an example, if the amplitude $|H(\omega)|$ of a certain filter is zero at $\omega = \omega_0$, then the output signal will be zero, meaning that the input signal with frequency ω_0 is blocked by this filter. For another example, if the phase $\phi(\omega)$ of a certain filter is negative at $\omega = \omega_0$, then the output signal will be lagged in time.

More generally, if the original input is expressed in the form of exponential Fourier series expansion with multiple frequencies ω_k:

$$x_j = \sum_{k=-\infty}^{\infty} c_k \exp(ij\omega_k),$$

then the output will be

$$y_j = \sum_{k=-\infty}^{\infty} c_k H(\omega_k) \exp(ij\omega_k)$$

$$= \sum_{k=-\infty}^{\infty} c_k |H(\omega_k)| \exp(i(j\omega_k + \phi(\omega_k))).$$

Similarly, the amplitude decides which frequency can pass through, and the phase function determines how lagging or leading is the output compared to the original input.

The amplitude or the gain function is used to define whether a filter is of low-pass or high-pass. A low-pass filter should have nonzero and significant amplitude for frequency range $0 \leq \omega \leq \omega_c$, while a high-pass filter should have nonzero and significant amplitude for frequency range $\omega_c \leq \omega$. We can also define a band-pass filter analogously if its amplitude is nonzero and significant between two non-zero frequency values.

For an ideal low-pass or high-pass filter, its amplitude would take the value one within the corresponding frequency range and zero otherwise. If we plot its amplitude versus different frequencies, the figure should look like a brick/rectangular. Note that the corresponding concept is the rectangular function in mathematics. Even though this property of cutting off undesirable frequencies sounds attractive, unfortunately this is impractical as real-life filters are not ideal. Thus, their amplitudes for most of the time is more like hump-shaped instead of brick-looking. This is called the leakage effect by engineers.

For instance, we take a look at the CMA in our first example and obtain its frequency response, namely the DTFT of the coefficients:

$$H(\omega) = \frac{1}{2M+1} \sum_{k=-M}^{M} \exp(-ik\omega) = \frac{1}{2M+1} \left[1 + 2 \sum_{k=1}^{M} \cos(k\omega) \right],$$

which is a real number.

The amplitude is plotted in Figure 5.3. We see that the amplitude is mainly nonzero for lower frequency values, and occasionally zero for certain nodal values. Hence the CMA is a low-pass filter. When it is acting on the original financial data, the high-frequency component should be filtered out, leaving the low-frequency component behind. The output will look like a smoothed version of the original data, as seen in Figure 5.1.

Since the frequency response of the CMA is a real number, its phase function is always zero. Thus no phase lag will be produced when processing a time series with the CMA. This is exactly what happened in Figure 5.1.

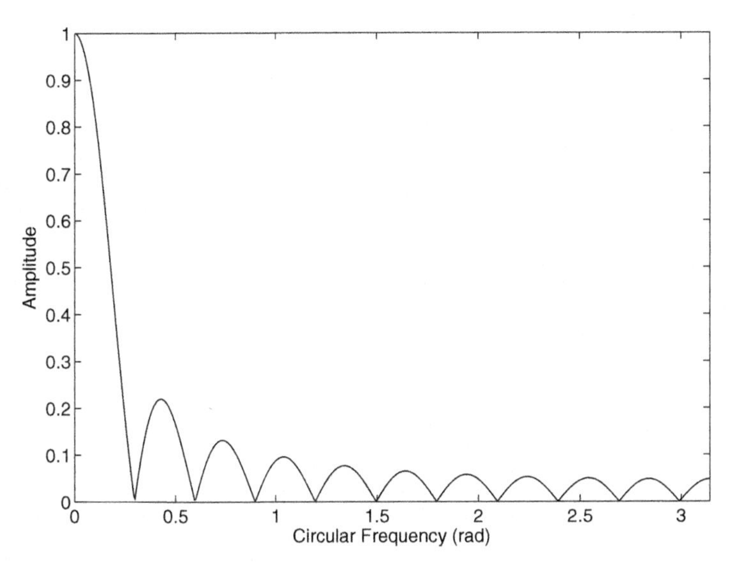

Figure 5.3: The amplitude of the frequency response of CMA_{10}, the CMA with $M = 10$

In fact, for any filter whose coefficients satisfy $h_k = h_{-k}$, we must have

$$H(\omega) = h_0 + 2\sum_{k=1}^{\infty} h_k \cos(k\omega),$$

which is a real number, and thus no phase shift happens. However, such filters are not practical because they make use of the future data. For practical purposes, we should consider the frequency response of a causal system

$$H(\omega) = \sum_{k=0}^{\infty} h_k \exp(-ik\omega).$$

In this case, $H(\omega)$ could be complex with a nonzero phase function. For example, we can derive the frequency response of an M-period SMA

$$H(\omega) = \frac{1}{M}\sum_{k=0}^{M-1} \exp(-ik\omega) = \frac{1}{M}\frac{1 - \exp(-iM\omega)}{1 - \exp(-i\omega)}.$$

In particular, its phase function is a linear function (Mak, 2006):

$$\phi(\omega) = -\frac{M-1}{2}\omega. \tag{5.3}$$

In Figure 5.4, we plot the amplitude and phase function of the frequency response of a SMA with $M = 20$. The amplitude has nonzero values mostly around the low-frequency range, with occasional zeros for certain nodal

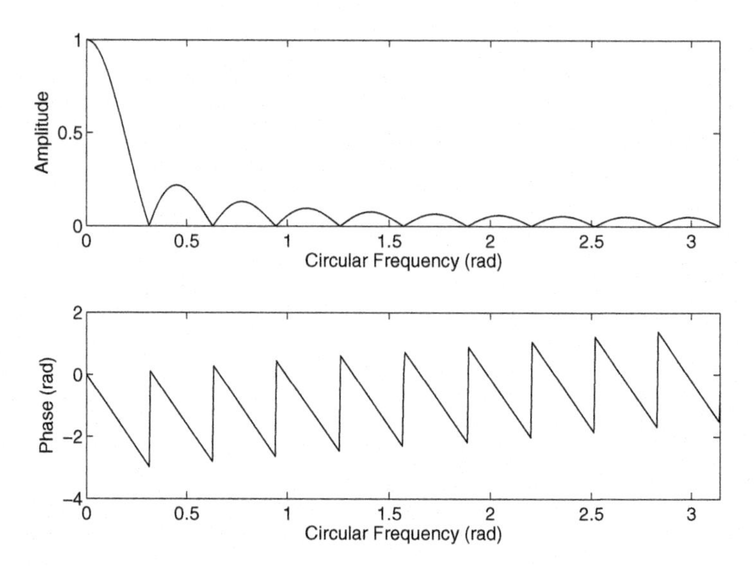

Figure 5.4: Frequency response of SMA_{20}, the SMA with $M = 20$

values. According to the definition, the 20-period SMA is a low-pass filter. Therefore, in Figure 5.2, we see that the SMA line is a filtered version of the original input data without its high-frequency fluctuations.

Note that when we plot the phase function of the SMA in the second panel of Figure 5.4, the jumps are caused by the built-in function `angle` in MATLAB as it returns values between $\pm\pi$. As shown previously in (5.3), the phase function should be linear and non-positive. Thus, the output produced by the SMA will be delayed in time, echoing the phenomenon in Figure 5.2.

5.3 Recursive Filters

In an M-period SMA, we see that the weights are equally distributed in the last M days. In order to emphasize the influences from the recent data, the M-period exponential moving average (EMA) has been proposed:

$$y_j = \alpha x_j + (1 - \alpha) y_{j-1}, \tag{5.4}$$

where $\alpha = 2/(M+1)$. For a 20-period EMA, the weight of the current data x_j is $2/(20+1) \approx 9.5\%$. Recall that for a 20-period SMA, the weights are all $1/20 = 5\%$.

To clearly see that more weights are put into the recent data and the weights fade away when time is further away from the present, we expand

y_j into the ordinary format of a linear filter:

$$\begin{aligned} y_j &= \alpha x_j + (1-\alpha)y_{j-1} = \alpha x_j + (1-\alpha)[\alpha x_{j-1} + (1-\alpha)y_{j-2}] \\ &= \alpha x_j + \alpha(1-\alpha)x_{j-1} + (1-\alpha)^2 y_{j-2} \\ &= \cdots = \alpha x_j + \alpha(1-\alpha)x_{j-1} + \cdots + \alpha(1-\alpha)^k x_{j-k} + (1-\alpha)^{k+1} y_{j-k-1} \\ &= \cdots. \end{aligned}$$

Note that when $M = 1$, we have $\alpha = 1$ and the output y_j simply equals to the input x_j. Therefore, we usually consider $M \geq 2$ and hence $\alpha < 1$. Under this scenario, we see that the ratio is heavier in recent data. Old data are not dropped out suddenly but slowly disappear because of the fading coefficients. The EMA keeps track of larger movements of the market because the output y_j becomes larger if the input x_j is larger. Also, the coefficients are exponentially decaying to zero, hence the name of the filter. For future usage, we also denote the M-period EMA by the following summation with infinite terms

$$\begin{aligned} y_j &= \alpha \sum_{k=0}^{\infty} (1-\alpha)^k x_{j-k} \\ &= \frac{2}{M+1} \sum_{k=0}^{\infty} \left(\frac{M-1}{M+1} \right)^k x_{j-k}. \end{aligned} \tag{5.5}$$

One useful feature of the EMA is that it requires little data storage, despite of its seemingly complicated definition. Only the current day input x_j and the most recently derived output y_{j-1} are needed. When we calculate a new value for the EMA, we simply update the old value using the new market data input. Another point is that the EMA requires a startup value for the recursive calculations, and this value is usually chosen as the SMA.

In Figure 5.5, we see the effect of a 20-period EMA on the Hang Seng Index in 2011. Similar to the SMA, the output signal is the smoothed version of the input signal and it lags in time. To explain these factors, we resort to the frequency response as usual. Note that we can always study the frequency response of a recursive filter after we have expanded it into the original linear filter, which has a format of infinite terms. However, we can also do it in an alternative way.

More generally, the EMA belongs to the class of recursive filters as defined in the following:

$$y_j = \sum_{k=-\infty}^{\infty} h_k x_{j-k} + \sum_{k=1}^{\infty} w_k y_{j-k}.$$

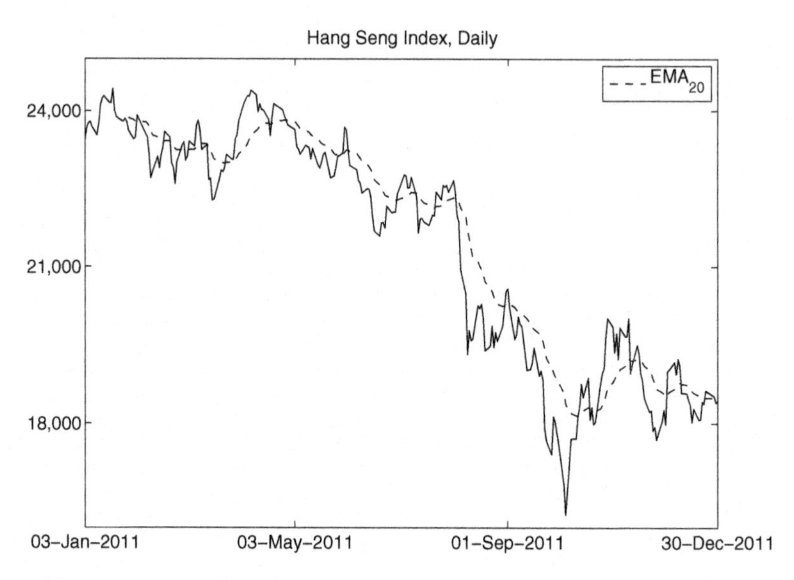

Figure 5.5: EMA with $M = 20$ applied on the Hang Seng Index in 2011

Note that the recursive filters only form a subclass of the linear filters, which is more obvious if we expand the recurring terms y_{j-k}'s repeatedly. The reason we write the filters in the recursive form is that we can study its frequency response more directly. Carrying out similar calculations using an eigenfunction $\mathbf{x} = \{x_j\} = \exp(ij\omega)$ as input data, we can show that the frequency response of a recursive filter is

$$H(\omega) = \frac{\sum_{k=-\infty}^{\infty} h_k \exp(-ik\omega)}{1 - \sum_{k=1}^{\infty} w_k \exp(-ik\omega)}.$$

Therefore, we can also study the amplitude and phase function in the frequency response of a recursive filter and explain its behaviors in time domain.

For example, the frequency response of an M-period EMA is given by

$$H(\omega) = \frac{\alpha}{1 - (1 - \alpha)\exp(-i\omega)},$$

where $\alpha = 2/(M + 1)$. In the following, we demonstrate the frequency response of an EMA using $M = 20$ and hence $\alpha = 2/(20+1)$. The amplitude and phase function of its frequency response are plotted in Figure 5.6. We observe that the amplitude of the EMA is less oscillating than the one

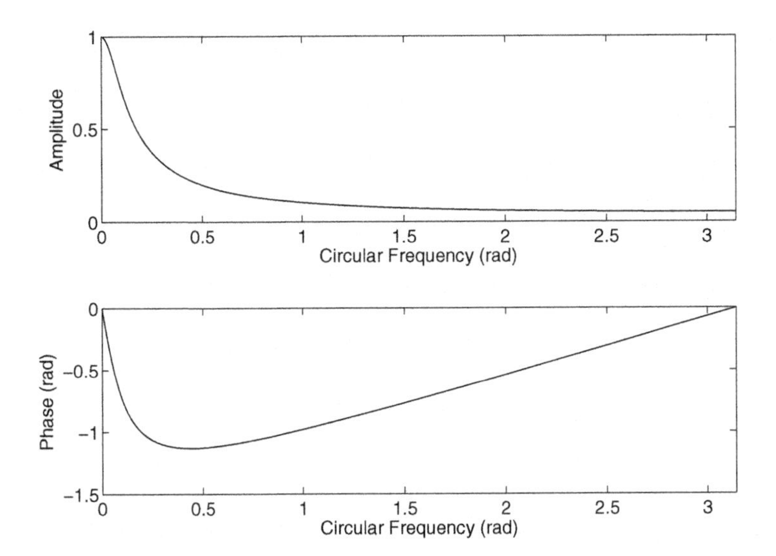

Figure 5.6: Frequency response of EMA_{20}, the EMA with $M = 20$

for SMA. Since the amplitude of the SMA is zero for certain frequency values, the SMA blocks out some of the lower frequency component as well. Thus, strictly speaking, it is not a low-pass filter. On the other hand, the amplitude of the EMA is smooth and lets all the low-frequency components pass through. All in all, the EMA is a better-behaved low-pass filter than the SMA.

For the phase function, we still see that it is non-positive, just like the one for SMA. Thus, the EMA results in a lagged output when it acts on a given signal.

5.4 Convolution Theorem

Traders sometimes combine linear filters for certain reasons, namely to consecutively apply two or more filters on the input signal. Such action is called the multistage filtering in signal processing. The final output can still be regarded as the filtered product of the original data via a combined linear filter. In time domain, the coefficients of this combined linear filter is simply the convolution of the coefficients of the original linear filters. We illustrate this property in the following example.

Suppose the given input signal is $\mathbf{x}_j$. If the first linear filter being applied is the 2-period SMA, then the first output will be

$$y_j = \frac{1}{2}(x_j + x_{j-1}).$$

Next we apply a 3-period SMA onto $\mathbf{y}_j$, then we obtain the second output

$$\begin{aligned} z_j &= \frac{1}{3}(y_j + y_{j-1} + y_{j-2}), \\ &= \frac{1}{3}\left[\frac{1}{2}(x_j + x_{j-1}) + \frac{1}{2}(x_{j-1} + x_{j-2}) + \frac{1}{2}(x_{j-2} + x_{j-3})\right], \\ &= \frac{1}{6}x_j + \frac{1}{3}x_{j-1} + \frac{1}{3}x_{j-2} + \frac{1}{6}x_{j-3}. \end{aligned}$$

Therefore, the second output $\mathbf{z}_j$ is the output of $\mathbf{x}_j$ via a new combined linear filter with filter coefficients

$$h_0 = \frac{1}{6}, \quad h_1 = \frac{1}{3}, \quad h_2 = \frac{1}{3}, \quad \text{and} \quad h_3 = \frac{1}{6}.$$

In fact, the coefficients of this new linear filter $\mathbf{h}$ is the convolution of the two original filter coefficients:

$$\mathbf{h} = \mathbf{h}_1 * \mathbf{h}_2 = \left(\frac{1}{2}, \frac{1}{2}\right) * \left(\frac{1}{3}, \frac{1}{3}, \frac{1}{3}\right) = \left(\frac{1}{6}, \frac{1}{3}, \frac{1}{3}, \frac{1}{6}\right).$$

Also note that such operation is commutative. Therefore, it does not matter which linear filter is applied first. The final output after two consecutive linear filters will be the same no matter what order you apply them.

In frequency domain, recall that the frequency response is simply the DTFT of the linear filter coefficients. This property, together with the famous convolution theorem [Oppenheim and Schafer (1989)], leads to the fact that the frequency response of a combined linear filter is simply the point-wise product of the two original frequency responses:

$$\begin{aligned} H(\omega) &= \text{DTFT}\{\mathbf{h}_1 * \mathbf{h}_2\}, \\ &= \text{DTFT}\{\mathbf{h}_1\} \cdot \text{DTFT}\{\mathbf{h}_2\}, \\ &= H_1(\omega) \cdot H_2(\omega). \end{aligned}$$

In Figure 5.7, we plot the frequency responses of EMA with $M = 10$ and $M = 20$ respectively, and also the frequency response of the two-stage filtering of these two EMAs. Since two low-pass filters are used consecutively, it is reasonable that the combined linear filter is a stronger low-pass filter making the data input smoother but more lagged in time.

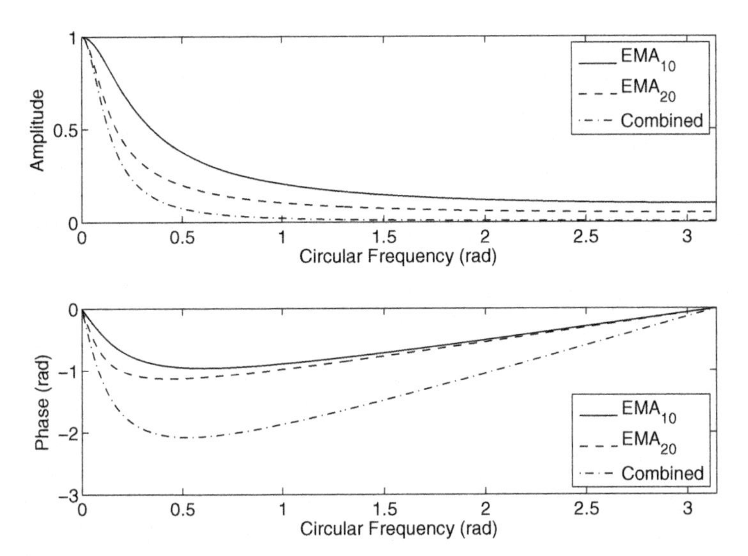

Figure 5.7: Frequency response of EMA with $M = 10$, $M = 20$ and their combination

5.5 Summary

This chapter concludes the basis of linear filters, including the definition of a linear filter, its frequency response, amplitude (gain function), phase function, and how these factors affect the processing of the original data. In this chapter, the examples we have considered are mainly low-pass filters. In the next chapter, we will focus more on band-pass and high-pass filters. We will also learn how to use linear filters to construct momentum indicators, essentially the tool to determine when to enter or exit the market.

Chapter 6

Momentum Indicators

In Chapter 5, we have introduced the linear filters to which many existing trading tools belong. Despite the observable behaviors in the time domain, it is beneficial to analyze linear filters in the frequency domain because amplitudes and phase shifts reveal their characteristics. In this chapter, we will discuss how to produce trading decisions using certain combination of linear filters. As long as such combination is linear, the overall filter is still linear. Therefore, it allows us to look into their frequency responses to figure out their properties.

In mathematics, one can tell the rises and falls of a smooth function by looking into its first-order derivative. When the first-order derivative is positive, the function rises. When the first-order derivative is negative, then the function falls. The idea can be borrowed for finding the market entry timing, but then we need to find the corresponding "first-order derivative" for processing price dynamics. In this context, we call such "first-order derivative" in the technical analysis field as the momentum indicator.

6.1 2-point Moving Difference

We first take a look at an indicator known as the 2-point moving difference

$$y_j = \frac{1}{2}(x_j - x_{j-1}),$$

where x_j is the closing price at time j. The trading strategy based on the 2-point moving difference is to keep track of its positivity every day. If the 2-point moving difference changes from negative to positive, then we enter the market and buy a share of the stock. On the other hand, if the 2-point moving difference changes from positive to negative, then we leave the market and sell the stock that we own, or short sell the stock if the market allows.

Note that the 2-point moving difference is simply known as the "momentum" among technical analysts. In mathematics, it is also known as the backward difference, an approximation to the first-order derivative. By studying its frequency response in Section 6.2, we will see that the 2-point moving difference is a high-pass filter with phase lead $\pi/2$ near $\omega = 0$. These characteristics can be found in other popular trading methods such as the crossover of two moving averages and the moving average convergence–divergence (MACD), which will be discussed in later sections. Intuitively, they are also approximations to the first-order derivative in filtering theory, just like the 2-point moving difference. We adopt the name "momentum indicators" for indicators which act as the approximation to the first-order derivative. Momentum indicators are capable of capturing the market's trend, just like how a first-order derivative determines the rise and fall of a function.

6.2 Momentum Indicators

We have mainly studied moving averages in Chapter 5. Their frequency responses show that they are low-pass filters. These moving averages resemble the smoothed versions of the original data and reflect its seasonalities and business cycle. In order to extract the changes or the dynamics of the market, we instead aim to filter out the slow-varying components. In particular, we employ the momentum indicators, a class of filters that behave like the first-order derivative of the input signal.

To figure out the characteristics of the 2-point moving difference, we consider its frequency response

$$H(\omega) = \frac{1}{2}[1 - \exp(-i\omega)].$$

We plot the frequency response of the 2-point moving difference in Figure 6.1.

From Figure 6.1, the 2-point moving difference is a high-pass filter, particularly with zero amplitude at $\omega = 0$. For the phase function, we observe that the phase starts from leading by $\pi/2$ near $\omega = 0$ and the lead is gradually decreasing as the frequency increases. Note that an ideal momentum indicator (or known as the first-order digital differentiator in signal processing) would have a phase lead of $\pi/2$ [Mak (2006)].

However, the high-pass feature of the 2-point moving difference is not wanted by technical analysts in the sense that its predictability is

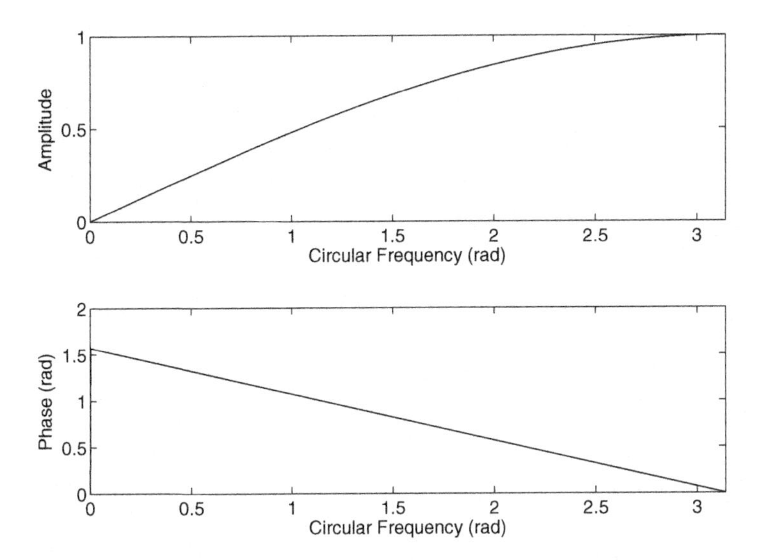

Figure 6.1: Frequency response of the 2-point moving difference

contaminated by the noise contained in a financial time series. The direct usage of 2-point moving difference is not very common because it generates a lot of buy and sell signals, leading to frequent trades. A general way to improve the indicator is to smooth the input signal beforehand, though from Chapter 5 we have seen that smoothing tools such as moving averages will bring phase lag to the original signal. In the forthcoming sections, we will focus on momentum indicators which bring about market entry timing, but at the same time try to get rid of the unwanted fluctuations.

6.3 Crossover of Two Moving Averages

Note that the 2-point moving difference can be formulated as the difference between the signal itself and its 2-period simple moving average (SMA):

$$y_j = x_j - \frac{1}{2}(x_j + x_{j-1}).$$

A momentum indicator can be regarded as the difference between the original signal and its lower frequency component. Thus, a generalization is to replace the 2-period SMA by other moving averages. First, we use the following notations to represent moving averages, e.g.,

$$y_j = \text{SMA}_M(\mathbf{x}_j) \equiv \frac{1}{M}(x_j + x_{j-1} + \cdots + x_{j-M+1})$$

or

$$y_j = \text{EMA}_M(\mathbf{x}_j) \equiv \alpha x_j + (1-\alpha)y_{j-1},$$

where $\alpha = 2/(M+1)$. Furthermore, let Id be the identity operator such that

$$\text{Id}(\mathbf{x}_j) = x_j.$$

Note that both Id and SMA_2 are linear filters. Then the 2-point moving difference can be written as

$$y_j = \text{Id}(\mathbf{x}_j) - \text{SMA}_2(\mathbf{x}_j) \equiv (\text{Id} - \text{SMA}_2)(\mathbf{x}_j),$$

where $(\text{Id} - \text{SMA}_2)$ can be regarded as a new linear filter acting on $\mathbf{x}_j$.

Replacing the 2-period SMA with other moving averages, we obtain new linear filters like

$$y_j = (\text{Id} - \text{SMA}_M)(\mathbf{x}_j) \quad \text{or} \quad y_j = (\text{Id} - \text{EMA}_M)(\mathbf{x}_j).$$

To see whether such kind of operations gives birth to momentum indicators, we check their behaviors in the frequency domain.

From Figure 6.2, it is clear that $(\text{Id} - \text{SMA}_{20})$ is a high-pass filter. For the phase function, it starts from leading by $\pi/2$ near $\omega = 0$ but the

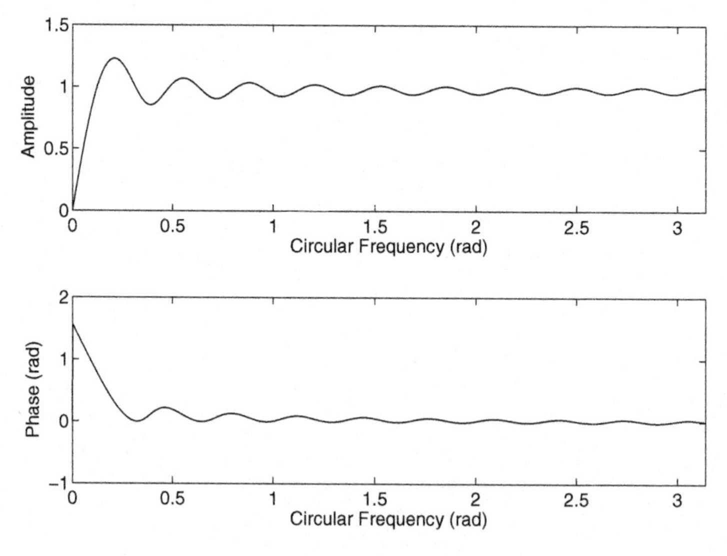

Figure 6.2: The frequency response of $(\text{Id} - \text{SMA}_{20})$

phase lead is dropping as frequency increases, resembling the behavior of the 2-point moving difference.

The positivity of ($\text{Id} - \text{SMA}_M$) or ($\text{Id} - \text{EMA}_M$) can be used to determine the timing of market entries. This is equivalent to checking when the data itself crosses its moving average from below or above. That is, if the input data is first below its moving average and then rises above it in the following day, meaning that the momentum indicator is changing from negative to positive, a buy signal is generated. On the other hand, when the input data drops below its moving average, meaning that the momentum indicator is changing from positive to negative, a (short-)selling signal is formed.

Figure 6.3 shows an example of Hang Seng Index from 2007 to 2011, where the 100-period SMA is applied. We check the crossovers of the data and the moving average for trading actions. The 100-period SMA appears as the lagged and smoothed version of the original data, as analyzed in Section 5.2. To illustrate the phase lead effect that exist in momentum indicators, we also plot the difference between the input data and its 100-period SMA within the same time domain. It can be clearly observed that their difference leads the original signal in time.

At points A and E, the data crosses the moving average from below, hence buying signals are generated. At points B and F, the data crosses

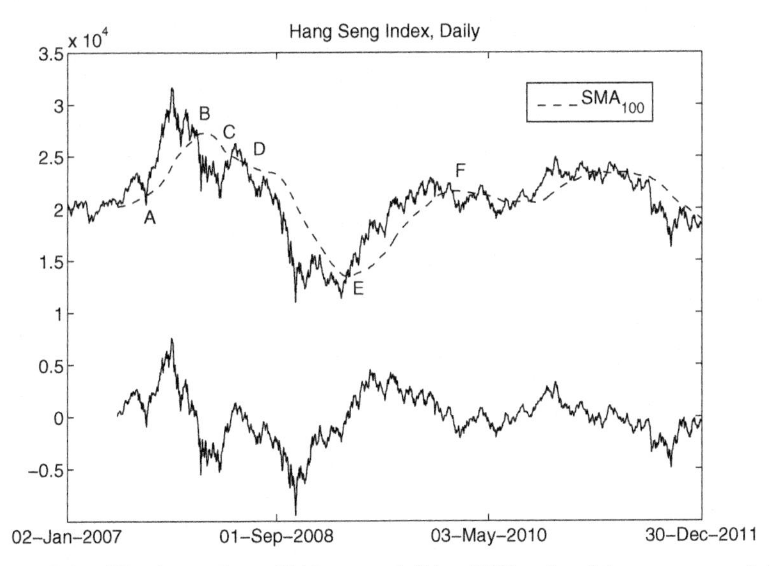

Figure 6.3: The input data, SMA_{100}, and ($\text{Id} - \text{SMA}_{100}$), while crossovers of data and SMA_{100} generate buying and selling signals

the moving average from above, hence selling signals are generated and the shares bought at points A and E respectively can be sold. However, signals can sometimes be false as shown in points C and D. Moreover, if the moving average in use has a shorter period, then more crossovers will take place, but at the same time more false signals will be produced as well. Another phenomenon worth noticing is that the market goes on a non-trending run in 2011, and the crossovers are more likely to tell investors to buy and sell at the same price level. On the other hand, while the market is trending from 2007 to 2009, the crossovers are more explicit and profitable.

Technical analysts also use the crossover of two moving averages as an indicator of whether to buy or sell in the market. In reality, this is the generalization of the previous case where we monitor the crossovers of the data and its moving average. It is because the data itself can be regarded as some kind of a 1-period moving average. For instance, the difference between two exponential moving averages (EMAs) is given by

$$y_j = \mathrm{EMA}_{M_1}(\mathbf{x}_j) - \mathrm{EMA}_{M_2}(\mathbf{x}_j), \quad M_1 < M_2,$$

where the first moving average with shorter period M_1 is known as the short moving average and the second one with longer period M_2 is known as the long moving average. We illustrate an example with $M_1 = 10$ and $M_2 = 50$.

From Figure 6.4, we see that the frequency response of ($\mathrm{EMA}_{10} - \mathrm{EMA}_{50}$) also behaves like the momentum indicator but with fewer high-frequency components passing through. Therefore, it is a band-pass filter. Also, the phase lead starts from $\pi/2$ near $\omega = 0$. Its positivity is usable for determining market entries, just like the 2-point moving difference. In other words, when the short moving average rises above the long moving average (momentum indicator changing from negative to positive), a buying signal is emitted. When the short one drops below the long one, a selling signal is formed.

An example of China Shanghai Composite Index from 2007 to 2011 is given in Figure 6.5. Since the two moving averages in use have longer periods, they capture the long-run movement and hence generate fewer crossovers.

6.4 Velocity Indicators

From the viewpoint of a Taylor series expansion, the 2-point moving difference is an approximation to the first-order derivative of order one,

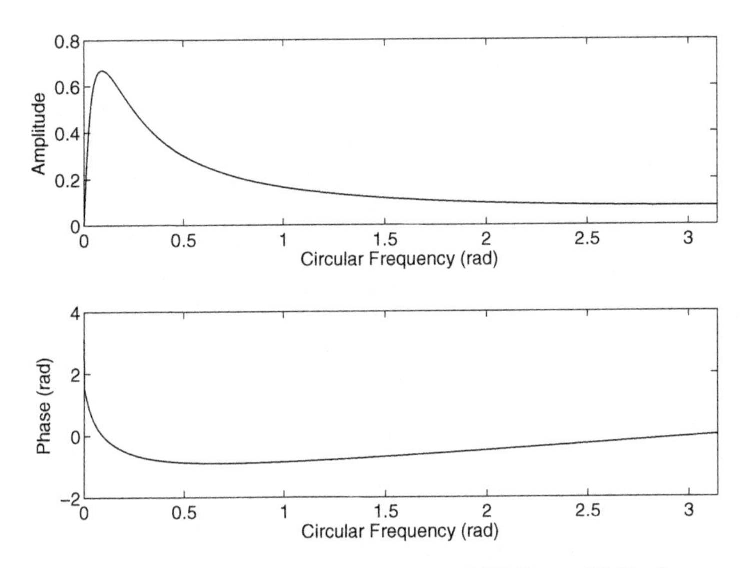

Figure 6.4: The frequency response of $(\text{EMA}_{10} - \text{EMA}_{50})$

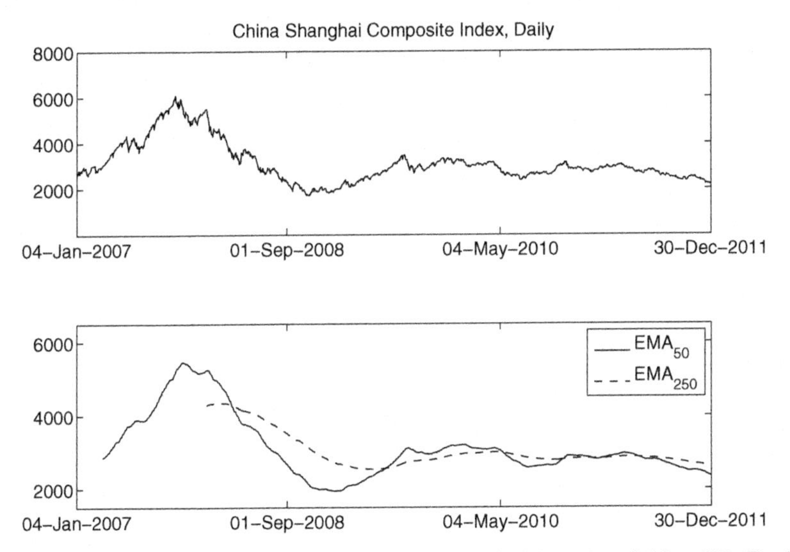

Figure 6.5: The crossover of two EMAs with periods $M = 50$ and $M = 250$. Buying and selling signals are generated from the crossovers

i.e., using only one neighbouring term for approximation. Therefore, it is possible to obtain higher-order approximations to the first-order derivative by maneuvering the Taylor series expansion. Mak (2006) names such momentum indicators, the velocity indicators. It is found out that the

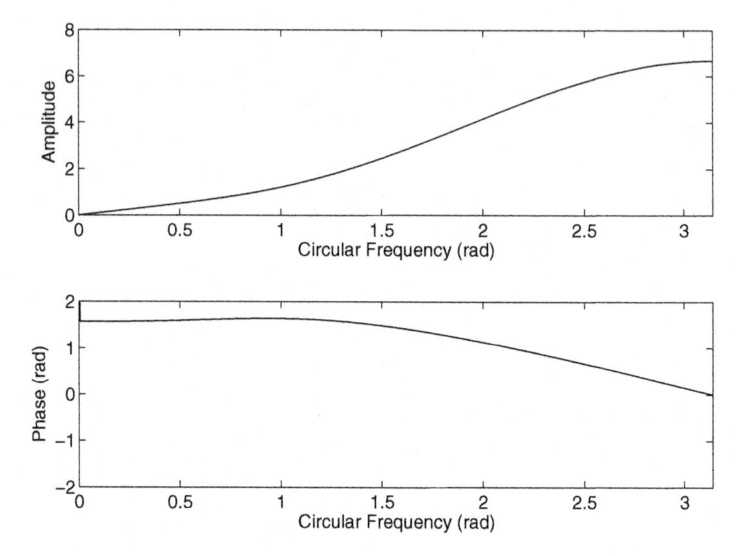

Figure 6.6: Frequency response of the cubic velocity indicator

2-point moving difference indeed is a velocity indicator of linear type. To move a step further, we define a cubic velocity indicator by

$$y_j = \mathrm{CVI}(\mathbf{x}_j) \equiv \frac{11}{6}x_j - 3x_{j-1} + \frac{3}{2}x_{j-2} - \frac{1}{3}x_{j-3}. \tag{6.1}$$

Here the word "cubic" means that the first-order derivative is approximated by three neighbouring terms (not an approximation containing a third-order derivative). The cubic velocity indicator is a linear filter and the four coefficients are derived by the method of undetermined coefficients [Mak (2003)].

From Figure 6.6, we see that the cubic velocity indicator is also a high-pass filter. Also from the phase function, the cubic velocity acts more like the ideal momentum indicator with phase lead $\pi/2$ for a wider frequency range. Thus, we can use the cubic velocity indicator to make trading decisions by checking its positivity.

6.5 MACD and Acceleration Indicators

MACD is a favorable technical analysis tool created by Gerald Appel. In its original form, it is defined as the difference of two EMAs with periods 12 and 26:

$$y_j = \mathrm{MACD}(\mathbf{x}_j) \equiv \mathrm{EMA}_{12}(\mathbf{x}_j) - \mathrm{EMA}_{26}(\mathbf{x}_j). \tag{6.2}$$

Since the MACD is the difference of two EMAs, we know from Section 6.3 that it is also a momentum indicator. Therefore, we can also design a trading strategy of buying and selling by checking the positivity of the MACD. When the MACD rises above 0, we enter the market, and when the MACD drops below 0, we leave the market.

Note that the parameters 12 and 26 are chosen by Gerald Appel but they can be changed according to different occasions. However, there is one rule to follow. The length of the first moving average must be shorter than the second one in order to produce a momentum indicator. The frequency response of the MACD is given in Figure 6.7.

Practitioners usually couple the MACD with the so-called signal line. It is the 9-period EMA of the MACD line:

$$y_j = \mathrm{EMA}_9(\mathrm{MACD}(\mathbf{x}_j)) \equiv (\mathrm{EMA}_9 \circ \mathrm{MACD})(\mathbf{x}_j).$$

Here the symbol $\circ$ represents the composition of the two linear filters. Thus, $(\mathrm{EMA}_9 \circ \mathrm{MACD})$ can be regarded as one linear filter. The coefficients of this combined linear filter is the convolution of the coefficients of the two original linear filters; see Section 5.4 for details. Recall that the EMA is a low-pass filter with phase lag. Hence the signal line is a lagged and smoothed version of the MACD line.

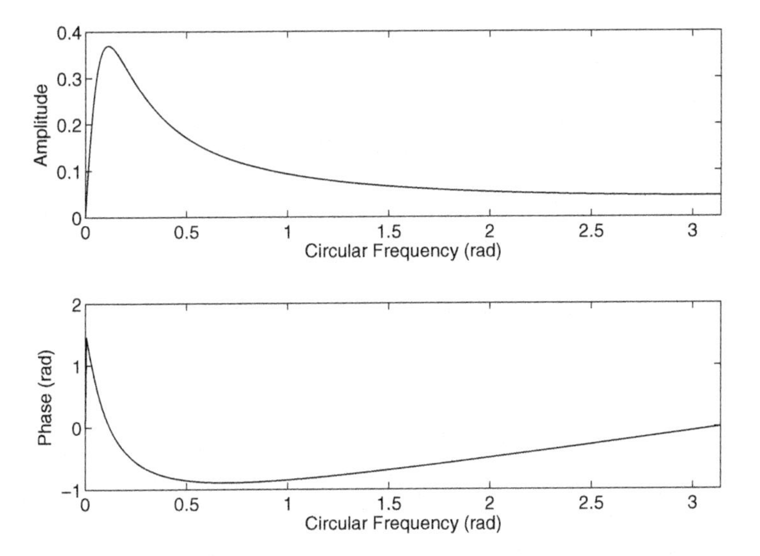

Figure 6.7: Frequency response of the MACD

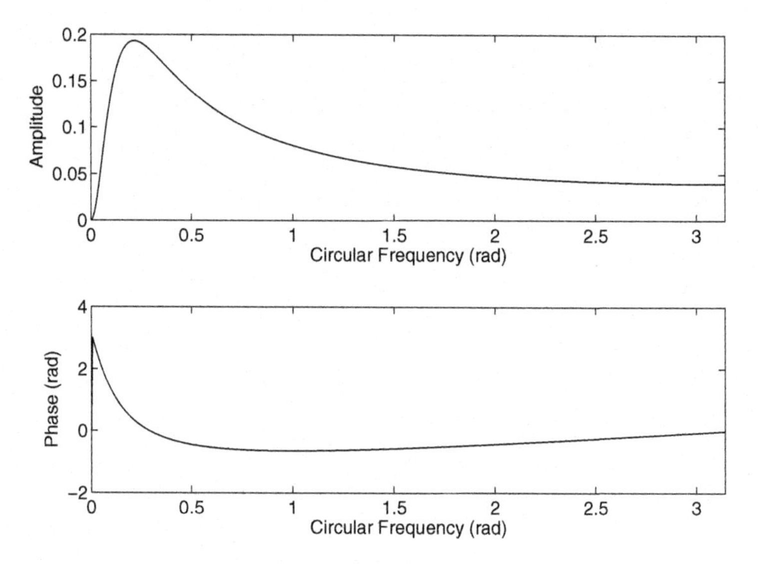

Figure 6.8: Frequency response of the MACD histogram

Another trading strategy based on MACD involves the crossovers of the MACD line and the signal line, i.e., to monitor the positivity of the following

$$\begin{aligned} y_j &= \text{MACD}(\mathbf{x}_j) - \text{EMA}_9(\text{MACD}(\mathbf{x}_j)) \\ &= [\text{MACD} \circ (\text{Id} - \text{EMA}_9)](\mathbf{x}_j), \end{aligned} \tag{6.3}$$

which is known as the MACD histogram. The frequency response of the MACD histogram is plotted in Figure 6.8.

From Figure 6.8, we see from its amplitude that MACD histogram is a band-pass filter. But the phase function looks different from the previous momentum indicators as the phase starts from leading by π instead of $\pi/2$ near $\omega = 0$. In reality, we can see from (6.3) that MACD is a momentum indicator, and $(\text{Id} - \text{EMA}_9)$ is also a momentum indicator. The MACD histogram is equivalent to two first-order derivatives being applied onto the original signal. In mathematics, this should represent a second-order differential operator. In this context, we will call them the acceleration indicators.

Mathematically, a positive second-order derivative does not imply a rise in the function. Therefore, the positivity of an acceleration indicator is not a detector for the rise and fall of the input data. Nevertheless, it can be used to determine the remaining dynamics in the market. For instance, a negative first-order derivative implies that the function is decreasing. However, at the same time if the second-order derivative is positive, the function possesses the power to rise and thus leads to a potential increase in the future.

Therefore, technical analysts use the crossovers of the MACD line and its signal line to obtain buying and selling signals, i.e., checking the positivity of the MACD histogram. They believe that when the MACD line crosses the signal line from below (MACD histogram becoming positive), there are still forces to push the market upwards, regardless of the positivity of the MACD line. When the MACD line crosses the signal line from above (MACD histogram becoming negative), the market is losing its dynamics and it is the time to sell or go short. The rules for such kind of aggressive trading method are as follows:

	MACD histogram > 0	MACD histogram < 0
MACD > 0	Buy	Sell
MACD < 0	Buy	Sell

We explain the different situations of using the MACD and the MACD histogram by an illustrative example given in Figure 6.9, where the Dow Jones Industrial Index in 2010 is given.

There are four kinds of crossovers in Figure 6.9. At point A, the MACD line crosses the signal line from above (MACD histogram changing from positive to negative) and the MACD value is positive. Equivalently, this

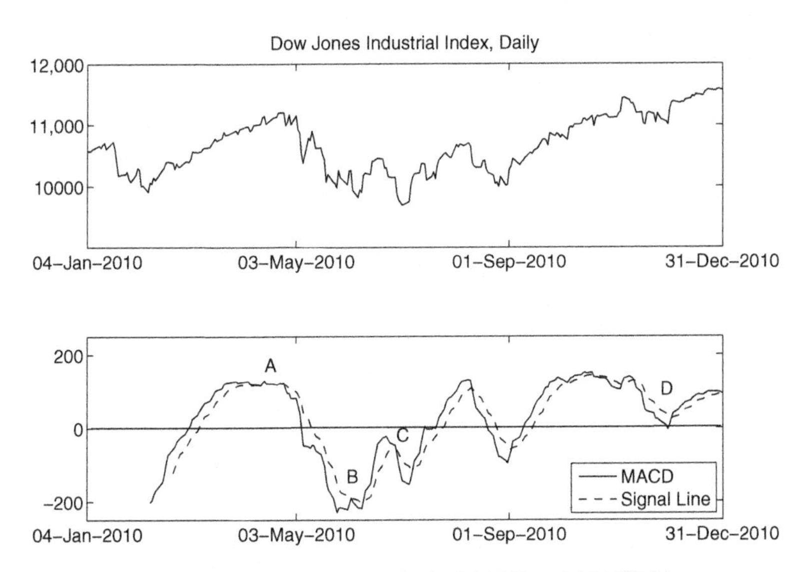

Figure 6.9: Trading strategy with the MACD and MACD histogram

is the case of a positive first-order derivative and a negative second-order derivative. Therefore, it implies that even though the market is rising, it is losing its power to rise further and hence it is time to leave the market at point A.

At point B, the MACD line crosses the signal line from below and the MACD value is negative. Equivalently, this is the case of a negative first-order derivative and a positive second-order derivative. It signifies the potential dynamics in the market even though it is currently on a down-trend movement. An aggressive trader will take this opportunity to enter the market.

At point C, the MACD line crosses the signal line from above and the MACD value is negative, similar to the case when both first- and second-order derivatives are negative. Thus, a selling signal is generated and it is a stronger selling signal than the one at point A. At point D, the MACD line crosses the signal line from below and the MACD value is positive, similar to the case when both first- and second-order derivatives are positive. Thus, a buying signal is generated and it is a stronger buying signal than the one at point B.

6.6 Summary

In this chapter, we have studied the important idea in deciding market entry timing, i.e., momentum indicator. It is the first-order derivative in the technical analysis world. Causal linear filters can be combined to produce momentum indicators, such as the crossover of two moving averages. Moreover, when combined with the acceleration indicator, the momentum indicator can provide more insights into the market trend. In the next chapter, we will study how to measure the performance of the trading methods in a statistical way.

Chapter 7

Profitability of Momentum Indicators[1]

One of the most common strategies used by long-term investors in the stock market is the buy-and-hold strategy. Indeed, even shareholders of Singapore Telecom are encouraged to adopt this strategy [Wong *et al.* (2005)], partly because it is easy to implement, and also because it does give a respectable return over a reasonably long period of time. The main disadvantage of such a method, however, is that the investor is exposed to the full volatility of the market. For an investor who is forced to liquidate in a bearish market, the loss can be quite substantial.

In this chapter, we examine whether such risks can be reduced by employing simple momentum indicators, such as the crossover of moving averages and the moving average convergence–divergence MACD. Furthermore, we evaluate the performance of these rules and check whether they outperform the buy-and-hold strategy.

7.1 Trading Methods Using Momentum Indicators

We start from the trading method of moving average crossover. The most widely-used moving average for this method is the simple moving average (SMA). It is just a simple average of the previous M-period prices $\{x_j\}$ when M is the duration of the moving average. Recall that we use the symbol SMA_M to denote the M-period SMA line or the corresponding linear operator. In the following, we specify its value $SMA_{M,j}$ at day j by

[1]This chapter is mainly based on Wong *et al.* (2005) by one of the authors with updated examples.

(see (5.1)):

$$SMA_{M,j} = \frac{1}{M}(x_j + x_{j-1} + \cdots + x_{j-M+1}).$$

The value $SMA_{M,j}$ is previously denoted as $\text{SMA}_M(\mathbf{x}_j)$, where the emphasis is on SMA_M being a linear operator on the input $\mathbf{x}_j$ up to time j. In the context that follows, we more often use the notation $SMA_{M,j}$ instead to focus on the specific value at time j.

A well-chosen moving average should reflect the trend. When the short moving average penetrates the long moving average, a change in trend may occur because the difference of two moving averages behaves like a momentum indicator. There are, of course, other types of moving average, like the weighted moving average and exponential moving average (EMA). However, we find in our empirical studies that when using the moving average crossover rule, the other moving averages are not necessarily better than the SMA. Therefore, for simplicity, we prefer the SMA for such crossover rule.

According to the moving average crossover rule, buy and sell signals are generated by comparing a short moving average and a long moving average. Buy (sell) signals are emitted when the short moving average rises above (falls below) the long moving average by a pre-specified percentage. This percentage forms a band around the long moving average. The band is used to reduce the number of times the investor would have to alternate between a long position (buying) and a short position (selling). It is because in general, the volatility of the moving average decreases with its duration. Hence for a short duration, one is more likely to experience a large number of false signals (called whipsaws). Conversely, it is also noted that the frequency of trades decreases with an increase in duration.

In this context, we test the technical trading rules on indices. Let $(x_1, x_2, \ldots, x_n)$ be an n-length realization of the index, where x_j is the index closing price on day j for $j = 1, 2, \ldots, n$. We classify each day as either being in a long position (buy), a short positive (sell) or a neutral position (no action). Suppose the moving average crossover rule is coupled with a $p\%$ band. Then we assume that a day j belongs to the buy period if

$$SMA_{M_1,j} > SMA_{M_2,j} \times \left(1 + \frac{p}{100}\right)$$

and a day j belongs to the sell period if

$$SMA_{M_1,j} < SMA_{M_2,j} \times \left(1 - \frac{p}{100}\right),$$

where $M_1 < M_2$. Such classification is used to measure the profitability of a trading method in Section 7.2.

For MACD, we first recall the EMA with duration M, and the indicating line is denoted by EMA_M. Its value $EMA_{M,j}$ at a specific day j is given by (see (5.5)):

$$EMA_{M,j} = \frac{2}{M+1}\sum_{k=0}^{\infty}\left(\frac{M-1}{M+1}\right)^k x_{j-k}.$$

At day j, the MACD value, denoted by $MACD_j$, is defined as the difference of $EMA_{12,j}$ and $EMA_{26,j}$ (see (6.2)):

$$MACD_j = EMA_{12,j} - EMA_{26,j}.$$

There are two types of trading methods involving the MACD. The first one, also considered as a more conservative trading method, is to simply use the positivity of MACD to decide market entries and exits. This idea originates from the fact that the MACD is the difference of two EMAs and thus behaves like a momentum indicator; see Section 6.5. Then we can also classify each trading day j into long or short period. A day j belongs to the buy period if

$$EMA_{12,j} > EMA_{26,j} \quad (MACD_j > 0).$$

A day j belongs to the sell period if

$$EMA_{12,j} < EMA_{26,j} \quad (MACD_j < 0).$$

The second type of trading method involves MACD and the MACD histogram and it is regarded as a more aggressive trading method. Recall that the MACD histogram, which will be denoted by MACDH, is the difference between the MACD line and its signal line. The signal line, which will be denoted by MACDS, is simply a 9-period EMA of the MACD line and its value at day j is recursively given by

$$MACDS_j = \frac{1}{5}\sum_{k=0}^{\infty}\left(\frac{4}{5}\right)^k MACD_{j-k}.$$

The MACD histogram is an acceleration indicator, or second-order derivative; see Section 6.5. This trading method makes use of the acceleration indicator to make market entries regardless of the momentum indicator. From the formula (6.3), we classify a day j as a part of the long period if

$$MACD_j > MACDS_j.$$

On the other hand, a day j belongs to the sell period if

$$MACD_j < MACDS_j.$$

In this context, we also test the effects of using the cubic velocity indicator, which is denoted by CVI, and its value at day j has the following formula (see (6.1)):

$$CVI_j = \frac{11}{6}x_j - 3x_{j-1} + \frac{3}{2}x_{j-2} - \frac{1}{3}x_{j-3}.$$

However, we notice in Section 6.4 that the CVI lets all high-frequency components pass through. The filtered output signal will still contain a lot of oscillations and thus it is not a stable indicator for market entries and exits. To eliminate the unwanted oscillating signals, we can apply an M-period EMA before applying the CVI. This combined strategy will generate an indicating line CVI_M and its value at day j is recursively given by

$$CVI_{M,j} = \frac{2}{M+1}\sum_{k=0}^{\infty}\left(\frac{M-1}{M+1}\right)^k CVI_{j-k}.$$

A day j is then classified as a buy day if $CVI_{M,j} > 0$ and a sell day if $CVI_{M,j} < 0$.

7.2 Performance Evaluation by Statistical Tests

Evaluation for the performance of different trading rules is done by using test statistics of return. The test statistics are used to determine whether the buy and sell signals give significantly profitable returns on the index, and assess the difference of the returns generated by the buy and sell signals. In statistics, the hypothesis testings can be used to assess if the means of two data groups are statistically different from each other in order to compute their means.

First, returns are defined as the natural logarithm of value relatives, which is similar to the arithmetic return for small values. We define the

1-day holding period return at day j as

$$r_j = \log\left(\frac{x_{j+1}}{x_j}\right), \quad j = 1, 2, \ldots, n-1,$$

where x_j is the closing price of the index at day j. We couple each daily closing price x_j with the 1-day holding period return r_j. Since the number of returns for the observation period is $n-1$, in the following, we let $N = n-1$ for convenience. In such case, the last day x_n is not coupled with any return since the future data is not available and we do not define the 1-day holding period return at day n. In summary, we have the following:

Day	1	2	$\cdots$	N	n
Price	x_1	x_2	$\cdots$	x_N	x_n
Return	r_1	r_2	$\cdots$	r_N	$-$

For example, if a trading rule shows that days 3, 4 are buy days and day 5 is the first sell day that comes afterwards, then the log-return of a long strategy (buying on day 3 and selling on day 5) would be:

$$\log\left(\frac{x_5}{x_3}\right) = \log\left(\frac{x_4}{x_3}\right) + \log\left(\frac{x_5}{x_4}\right) = r_3 + r_4,$$

i.e., one only needs to compute the sum of r_3 and r_4 once the buy days are determined.

To compare the various trading strategies with the static strategy, namely the buy-and-hold strategy, we use its mean return as the benchmark measure

$$\hat{\mu} = \frac{1}{N}\sum_{j=1}^{N} r_j = \frac{1}{N}\log\left(\frac{x_n}{x_1}\right),$$

which is proportional to the log-difference of the index between the first closing price x_1 and the last closing price x_n, thus reflecting the return from using the buy-and-hold strategy. In addition, we also calculate the mean variance for later usage

$$\hat{\sigma}^2 = \frac{1}{N}\sum_{j=1}^{N}(r_j - \hat{\mu})^2.$$

We first classify each day in the whole sample as either a buy or sell day according to the trading rules introduced in Section 7.2. If a day does not belong to the buy or sell period, it is classified as a neutral day. This happens when a $p\%$ band is incorporated. Note that there is no need to classify the day n because r_n is not available. Hence all the buy days, denoted by a set of integers Ω_{b}, will form a subset of $\{1, 2, \ldots, N\}$. And respectively all the sell days, denoted by a set of integers Ω_{s}, will form a subset of $\{1, 2, \ldots, N\}$. Furthermore, we let N_{b} and N_{s} be the size of the two sets Ω_{b} and Ω_{s} respectively. These two numbers represent how many days an investor is staying in the buy period or sell period.

The mean return and variance conditional on a buy signal can be written as

$$\hat{\mu}_{\mathrm{b}} = \frac{1}{N_{\mathrm{b}}} \sum_{j \in \Omega_{\mathrm{b}}} r_j$$

$$\hat{\sigma}_{\mathrm{b}}^2 = \frac{1}{N_{\mathrm{b}}} \sum_{j \in \Omega_{\mathrm{b}}} (r_j - \hat{\mu}_{\mathrm{b}})^2.$$

Similarly, the mean return and variance conditional on a sell signal are denoted by

$$\hat{\mu}_{\mathrm{s}} = \frac{1}{N_{\mathrm{s}}} \sum_{j \in \Omega_{\mathrm{s}}} r_j,$$

$$\hat{\sigma}_{\mathrm{s}}^2 = \frac{1}{N_{\mathrm{s}}} \sum_{j \in \Omega_{\mathrm{s}}} (r_j - \hat{\mu}_{\mathrm{s}})^2.$$

After an investor decides which trading rule to use, the set of buy days and sell days Ω_{b} and Ω_{s} are determined. Then they are used to compute the mean daily return and variance conditional on these two sets. The mean return $\hat{\mu}_{\mathrm{b}}$ is the average daily return for the long strategy generated by the trading strategy and $\hat{\mu}_{\mathrm{s}}$ is the average daily return for the short strategy generated by the trading method. Meanwhile, $\hat{\sigma}_{\mathrm{b}}$ and $\hat{\sigma}_{\mathrm{s}}$ are the standard deviations of the daily returns generated by the buy signals and sell signals respectively.

To value the significance of the return generated by a trading strategy, we introduce the hypothesis

$$H_{01} : \mu_{\mathrm{b}} = \mu \text{ against } H_{11} : \mu_{\mathrm{b}} > \mu$$

to test whether the return is profitable for the long strategy compared with the buy-and-hold strategy. Similarly, the hypothesis

$$H_{02} : \mu_{\mathrm{s}} = \mu \text{ against } H_{12} : \mu_{\mathrm{s}} < \mu$$

is used to test whether the return is profitable for the sell signal compared with the buy-and-hold strategy. Alternatively, since we expect that the returns will be positive for the buy signal and negative for the sell signal, we can also perform tests on $\mu_b > 0$ and $\mu_s < 0$ to see whether the buy and sell signals generated by the trading rules yield significantly positive returns. Owing to a limited context, we only report the results of hypotheses H_{01} against H_{11} and H_{02} against H_{12} as they are commonly used in literature like Brock *et al.* (1992); Kwon and Kish (2002); Metghalchi *et al.* (2008).

For the long strategy, the test statistic for the buys is [Kwon and Kish (2002)]:

$$T_b = \frac{\hat{\mu}_b - \hat{\mu}}{\sqrt{\hat{\sigma}_b^2/N_b + \hat{\sigma}^2/N}}.$$

The test statistic for the sells is [Kwon and Kish (2002)]:

$$T_s = \frac{\hat{\mu}_s - \hat{\mu}}{\sqrt{\hat{\sigma}_s^2/N_s + \hat{\sigma}^2/N}}.$$

Both T_b and T_s are used as test statistics in a one-tailed hypothesis test. We note that the above hypotheses are used to test whether the return generated by the trading method in a long position and in a short position is significantly more profitable than that by the buy-and-hold strategy. Furthermore, we do not assume the standard deviation of the daily returns for the periods generated by the buy signals or sell signals to be equal to the standard deviation of the daily returns for the entire period. We remark that such assumption is used when an estimated standard deviation for the entire period is involved in the tests [Brock *et al.* (1992)].

It is often argued that the daily returns are not independent and identically distributed (i.i.d.) as normal. The violation of the normality assumption and independence assumption for daily returns is well known [Fama (1965); Fama and French (1988); Conrad and Kaul (1988); Lo and MacKinlay (1990)]. However, the test statistics will be approximately distributed as $\mathcal{N}(0, 1)$ if $\hat{\mu} = 0$. As n is usually large, it is not necessary to impose the normality assumption as the test statistics will still approach the standard normal distribution by the central limit theorem.

We also report the mean return difference, also known as the buy–sell spread. The hypothesis

$$H_{03} : \mu_b = \mu_s \text{ against } H_{13} : \mu_b \neq \mu_s$$

is used to test whether the long strategy is significantly different from the short strategy. The test statistic for the buy–sell difference is [Kwon and

Kish (2002)]

$$T_{\rm bs} = \frac{\hat{\mu}_{\rm b} - \hat{\mu}_{\rm s}}{\sqrt{\hat{\sigma}_{\rm b}^2/N_{\rm b} + \hat{\sigma}_{\rm s}^2/N_{\rm s}}},$$

and $T_{\rm bs}$ is used as the test statistic in a two-tailed hypothesis test. Once again, we can alternatively test whether a combined strategy of going long and short (when short selling is allowed) will yield significant positive returns by testing $\mu_{\rm b} - \mu_{\rm s} > 0$. However, owing to a limited context, we only report the results of the hypothesis H_{03} against H_{13} as it is used more in literature like Brock *et al.* (1992); Kwon and Kish (2002); Metghalchi *et al.* (2008).

To take into account the transaction costs, we also report the break-even costs in the numerical studies [Bessembinder and Chan (1995)]:

$$C = \frac{N_{\rm b}\hat{\mu}_{\rm b} - N_{\rm s}\hat{\mu}_{\rm s}}{n_{\rm b} + n_{\rm s}},$$

where $n_{\rm b}$ and $n_{\rm s}$ are the number of days on which the buy and sell signals are initially generated. The number is the percentage round-trip costs that eliminate the additional return from technical trading, i.e., the larger the value of C, the higher transaction costs a trading method can allow.

7.3 Evaluation Results

We employ the daily Hang Seng Index and China Shanghai Composite Index extracted from Yahoo! Finance for the period from January 2007 to December 2011. It is a bull run in 2007 before the financial crisis in 2008, which is a bear market. From 2009 onwards, it can be considered as a mixed market. The test statistics will be applied to test whether the buy and sell signals generated by various trading methods produce significant profits and surpass the buy-and-hold strategy in these two markets.

For the moving average crossover, we use the symbol $(M_1, M_2, p\%)$ to represent a trading strategy, where M_1 is the duration of the short moving average, M_2 is the duration of the long moving average, and a $p\%$ band is included to envelop the long moving average. In this study, the following ten trading rules [Brock *et al.* (1992)] are used: $(1, 50, 0\%)$, $(1, 50, 1\%)$, $(1, 150, 0\%)$, $(1, 150, 1\%)$, $(5, 150, 0\%)$, $(5, 150, 1\%)$, $(1, 200, 0\%)$, $(1, 200, 1\%)$, $(2, 200, 0\%)$, and $(2, 200, 1\%)$.

For the two trading strategies involving MACD, we use the symbol "MACD" to represent the basic MACD trading method, namely to enter and exit the market by judging the positivity of the MACD indicator.

We use the symbol "MACDH" to denote the advanced MACD trading method, namely to check the crossovers of MACD and its signal line for timing of market entries and exits.

For the velocity indicator [Mak (2006)], we first smooth the input data by an M-period EMA, then apply the CVI onto the smoothed data. We use the symbol "CVI_M" to denote such a combined trading strategy. For a complete roundup of all these trading methods, readers are referred to Section 7.2.

Tables 7.1 and 7.2 report the mean return $\hat{\mu}_{\text{b}}$ produced by the long strategy with the corresponding test statistic T_{b} (placed underneath $\hat{\mu}_{\text{b}}$ in the tables), the mean return $\hat{\mu}_{\text{s}}$ produced by the short strategy with the corresponding test statistic T_{s} (placed underneath $\hat{\mu}_{\text{s}}$ in the tables), the mean return difference $\hat{\mu}_{\text{b}} - \hat{\mu}_{\text{s}}$ with the corresponding test statistic T_{bs} (placed underneath $\hat{\mu}_{\text{b}} - \hat{\mu}_{\text{s}}$ in the tables), and the break-even costs C. In addition, the number of buy days and sell days generated by the trading methods, N_{b} and N_{s} respectively, are reported in the tables.

In Table 7.1, we find that $\hat{\mu}_{\text{b}}$'s, the mean daily returns for the long strategy are all positive, ranging from 0.010% to 0.085%. But the significant levels of T_{b} between the average daily returns generated by any momentum indicator and the average daily returns generated by the buy-and-hold strategy are not high. These results are not convincing enough for us to reject H_{01}, i.e., to conclude that the returns generated by holding the long strategy for any trading rule in the momentum indicator family perform significantly different from the buy-and-hold strategy for Hang Seng Index.

Table 7.1 also shows the performance of the short strategy generated by the trading rules from the momentum indicator family for Hang Seng Index. Their mean returns $\hat{\mu}_{\text{s}}$'s are all negative, but the significant levels of T_{s} between the average daily returns generated by the short strategy of any trading rule in the momentum indicator family and the average daily returns generated by the buy-and-hold strategy are not high. Therefore, these results are also not convincing enough for us to reject H_{02}, i.e., to conclude that returns generated by the short strategy from the trading rules in the momentum indicator family are significantly different from the buy-and-hold strategy in Hong Kong's stock market.

Break-even costs reflect the balance between the profitability of the trading rule and how many times an investor needs to enter and exit the market while executing this trading rule. If a trading rule generates high profits while not requiring frequent trades, then the break-even cost will be a larger number, because the larger the break-even cost, the higher transaction cost a trading rule can tolerate. For instance, in Table 7.1, the

Table 7.1: Standard test results and break-even costs for Hang Seng Index from 2007 to 2011, with momentum indicators

Rule	N_b	N_s	$\hat{\mu}_b$ (T_b)	$\hat{\mu}_s$ (T_s)	$\hat{\mu}_b - \hat{\mu}_s$ (T_{bs})	C
$(1, 50, 0\%)$	643	590	0.00049 (0.66)	−0.00070 (−0.53)	0.00120 (1.00)	0.0105
$(1, 50, 1\%)$	579	513	0.00038 (0.51)	−0.00063 (−0.43)	0.00101 (0.76)	0.0062
$(1, 150, 0\%)$	652	581	0.00056 (0.77)	−0.00080 (−0.60)	0.00136 (1.13)	0.0245
$(1, 150, 1\%)$	626	545	0.00070 (0.92)	−0.00087 (−0.63)	0.00157 (1.23)	0.0285
$(5, 150, 0\%)$	654	579	0.00085 (1.10)	−0.00113 (−0.87)	0.00198 (1.63)	0.1007
$(5, 150, 1\%)$	634	543	0.00082 (1.05)	−0.00096 (−0.70)	0.00177 (1.40)	0.0648
$(1, 200, 0\%)$	710	523	0.00020 (0.35)	−0.00046 (−0.30)	0.00067 (0.52)	0.0102
$(1, 200, 1\%)$	669	502	0.00020 (0.34)	−0.00061 (−0.40)	0.00081 (0.61)	0.0116
$(2, 200, 0\%)$	708	525	0.00020 (0.34)	−0.00046 (−0.29)	0.00066 (0.51)	0.0136
$(2, 200, 1\%)$	666	498	0.00022 (0.36)	−0.00052 (−0.34)	0.00074 (0.56)	0.0156
MACD	643	590	0.00014 (0.25)	−0.00032 (−0.20)	0.00045 (0.38)	0.0066
MACDH	594	639	0.00010 (0.19)	−0.00024 (−0.15)	0.00034 (0.29)	0.0020
CVI_{49}	644	589	0.00030 (0.44)	−0.00049 (−0.34)	0.00078 (0.65)	0.0021
CVI_{149}	716	517	0.00042 (0.62)	−0.00077 (−0.52)	0.00119 (0.91)	0.0062
CVI_{199}	756	477	0.00013 (0.26)	−0.00041 (−0.24)	0.00054 (0.39)	0.0035

moving average crossover rule $(5, 150, 0\%)$ is performing well with mean return 0.085%, or 21.42% annually. Since the two moving averages have comparatively longer durations, the trades do not take place so frequently as other moving average crossover rules like $(1, 50, 0\%)$. Hence, the break-even cost for $(5, 150, 0\%)$ is relatively high at 10.07%.

Overall, the simple momentum indicators do not generate significant profits in the Hong Kong stock market. This may be consistent with the weak-form efficient-market hypothesis proposed by Fama (1970), which states that a market is efficient if it is impossible for investors to make abnormal returns by using any publicly known investment strategies.

Table 7.2: Standard test results and break-even costs for China Shanghai Composite Index from 2007 to 2011, with momentum indicators

Rule	N_b	N_s	$\hat{\mu}_b$ (T_b)	$\hat{\mu}_s$ (T_s)	$\hat{\mu}_b - \hat{\mu}_s$ (T_{bs})	C
(1, 50, 0%)	589	626	0.00135 (1.58)	−0.00161 (−1.38)	0.00296 (2.57)	0.0311
(1, 50, 1%)	545	576	0.00140 (1.58)	−0.00141 (−1.14)	0.00282 (2.31)	0.0277
(1, 150, 0%)	633	582	0.00126 (1.50)	−0.00174 (−1.48)	0.00300 (2.58)	0.1006
(1, 150, 1%)	618	558	0.00114 (1.36)	−0.00182 (−1.52)	0.00296 (2.47)	0.0661
(5, 150, 0%)	637	578	0.00105 (1.27)	−0.00152 (−1.28)	0.00257 (2.21)	0.1290
(5, 150, 1%)	624	555	0.00111 (1.32)	−0.00176 (−1.48)	0.00287 (2.41)	0.1390
(1, 200, 0%)	634	581	0.00090 (1.13)	−0.00135 (−1.11)	0.00225 (1.93)	0.0564
(1, 200, 1%)	606	563	0.00100 (1.21)	−0.00139 (−1.13)	0.00239 (2.00)	0.0514
(2, 200, 0%)	632	583	0.00089 (1.11)	−0.00133 (−1.09)	0.00221 (1.90)	0.0667
(2, 200, 1%)	607	561	0.00101 (1.22)	−0.00147 (−1.20)	0.00247 (2.06)	0.0717
MACD	606	609	0.00083 (1.04)	−0.00117 (−0.95)	0.00200 (1.72)	0.0379
MACDH	642	573	0.00036 (0.57)	−0.00078 (−0.56)	0.00114 (0.97)	0.0068
CVI_{49}	593	622	0.00120 (1.45)	−0.00148 (−1.24)	0.00268 (2.32)	0.0101
CVI_{149}	608	607	0.00085 (1.06)	−0.00119 (−0.97)	0.00204 (1.76)	0.0132
CVI_{199}	613	602	0.00077 (0.97)	−0.00113 (−0.92)	0.00190 (1.64)	0.0134

Under this assumption, the Hong Kong stock market is an efficient one in the period from 2007 to 2011, partly due to popularization of online trading and the reduction in trading costs.

In Table 7.2, the same set of trading rules in the family of momentum indicators is applied to the China Shanghai Composite Index, in the period from 2007 to 2011. Most of the results are similar to those in Table 7.1, but the significant levels of T_b, T_s and T_{bs} are more obvious. This is probably due to legal person ownership and insider trading [Tian *et al.* (2002)] existing in the Chinese stock market. Therefore, the Chinese stock market is a more inefficient one and technical indicators can be used to obtain greater profits.

The results from Tables 7.1 and 7.2 have provided important insights into the performance of different trading strategies, where both the average daily returns from the long and short strategies and their differences from the average daily returns generated by the buy-and-hold strategy are clearly presented.

7.4 Summary

In our empirical studies, we find that the momentum indicators can determine the timing of market entries, and generate a considerable amount of profits when the market is inefficient. With the reduction of transaction cost and the allowance of short-selling, individual or institutional traders can furthermore gain greater profits, even by applying basic technical indicators like the crossover of two moving averages with different lengths. In Chapters 8 and 9, we will move further from the basic filtering theory and derive trading methods based on wavelet decomposition and empirical mode decomposition.

Chapter 8

Wavelets and Technical Analysis

Wavelets are a famous family of linear filters and a common tool for non-stationary time series. The wavelet analysis is able to handle cases that classic Fourier analysis cannot. The extensive studies in Gençay *et al.* (2001) show that wavelet multiresolution analysis (MRA) has applications such as seasonality filtering, denoising, and identifying structural breaks, etc., for financial or economical purposes.

In this chapter, we focus on combining wavelets and technical analysis to design a trading method similar to the moving average crossover rule. The idea is to apply the two major concepts in wavelet analysis, the father and mother wavelets. The father wavelet, also known as the scaling function, acts like a low-pass filter like the moving averages. The mother wavelet, on the other hand, can be scaled and dilated to generate different band-pass filters. Momentum indicators can then be constructed using these father and mother wavelets. However, the father and mother wavelets are usually not used individually as linear filters. They are important parts in the more advanced wavelet multiresolution analysis [Mallet (2008)], in which an input signal is decomposed into components with different frequencies. Afterwards, we can retrieve smoothed versions of the data by changing the number of terms being summed in the decomposed result. Then, a wavelet-based technical trading method is designed using the components derived from the multilevel wavelet decomposition.

8.1 One-Dimensional Continuous-time Wavelet Functions

A (mother) wavelet $\psi(t)$ is a function of time t that satisfies the wavelet *admissibility condition* [Gençay *et al.* (2001); Mallet (2008)]

$$C_\psi = \int_0^\infty \frac{|\Psi(\omega)|}{\omega} d\omega < \infty,$$

where $\Psi(\omega)$ is the Fourier transform of the wavelet $\psi(t)$. The admissibility condition implies that

$$\int_{-\infty}^{\infty} \psi(t)dt = 0. \tag{8.1}$$

Wavelets also satisfy another property, i.e., having unit energy

$$\int_{-\infty}^{\infty} |\psi(t)|^2 dt = 1. \tag{8.2}$$

Using the mother wavelet $\psi(t)$, we can next define the daughter wavelets with a dilation parameter a and a translation parameter b

$$\psi_{a,b}(t) = \frac{1}{\sqrt{a}}\psi\left(\frac{t-b}{a}\right).$$

By sampling the mother wavelet $\psi(t)$ at different scales a and b, we obtain a set of filters which represent time and frequency at different scale. To complete the wavelet analysis, the so-called father wavelet, or scaling function is introduced [Mallet (1989)]. It is usually denoted by $\phi(t)$ and satisfies the following condition

$$\int_{-\infty}^{\infty} \phi(t)dt = 1.$$

The scaling function acts like a low-pass filter because it integrates to 1, similar to the case in which moving average weights also add up to 1. On the other hand, we see that the mother wavelet integrates to 0 from (8.1). As a result, the daughter wavelets (mother wavelet included) act as band-pass filters [Mallet (2008); Gençay *et al.* (2001)]. These wavelets altogether form a family of filters in wavelet analysis.

To give a classic example, we take a look at the Haar wavelets proposed by Alfréd Haar in 1910 [Haar (1910)]. Haar wavelets are the simplest type of wavelets and they form the first known wavelet basis. The Haar scaling function or father wavelet is given by

$$\phi(t) = \begin{cases} 1, & 0 \le t < 1, \\ 0, & \text{otherwise.} \end{cases}$$

The Haar mother wavelet is given by

$$\psi(t) = \begin{cases} 1, & 0 \le t < \frac{1}{2}, \\ -1, & \frac{1}{2} \le t < 1, \\ 0, & \text{otherwise.} \end{cases}$$

The Haar daughter wavelets with dilation and translation parameters a and b are

$$\psi_{a,b}(t) = \begin{cases} \frac{1}{\sqrt{a}}, & b \le t < b + \frac{a}{2}, \\ -\frac{1}{\sqrt{a}}, & b + \frac{a}{2} \le t < b + a, \\ 0, & \text{otherwise.} \end{cases}$$

8.2 One-Dimensional Discrete Wavelet Transform

Let $x(t)$ denote an input signal. The continuous wavelet transform (CWT) is a function of two variables $W(a, b)$ with respect to the input $x(t)$:

$$W(a, b) = \int_{-\infty}^{\infty} x(t)\psi_{a,b}(t)dt,$$

where $\psi_{a,b}(t)$ are the daughter wavelets with dilation and translation parameters a and b. However, there are infinitely many choices of dilation and translation. This leads to the study of discrete wavelets for discrete-time signals in this section. In practice, we choose the so-called *dyadic* (fractional powers of two) dilation and translation parameters by

$$a = 2^{-j} \quad \text{and} \quad b = k2^{-j},$$

where both j and k are integers. Then we obtain the discrete version of daughter wavelets

$$\psi_{j,k}(t) = 2^{j/2}\psi(2^j t - k)$$

and the corresponding CWT is reduced to the case of discrete wavelet transform (DWT). Under the discrete setting, the mother wavelet is represented as M-length discrete wavelet filter whose coefficients $(h_0, h_1, \ldots, h_{M-1})$ satisfy

$$\sum_{k=0}^{M-1} h_k = 0 \quad \text{and} \quad \sum_{k=0}^{M-1} h_k^2 = 1.$$

The above conditions correspond to the two conditions (8.1) and (8.2) in the continuous case. For the scaling function, it is a discrete low-pass filter whose coefficients satisfy the quadrature mirror filter relationship [Gençay *et al.* (2001)]

$$g_k = (-1)^{k+1} h_{M-1-k}, \quad k = 0, 1, \ldots, M-1,$$

such that the orthogonality among wavelets holds [Mallet (2008)].

For instance, the Haar wavelet filter corresponds to the case $M = 2$ with

$$h_0 = \frac{1}{\sqrt{2}}, \quad h_1 = -\frac{1}{\sqrt{2}}, \quad g_0 = \frac{1}{\sqrt{2}}, \quad \text{and} \quad g_1 = \frac{1}{\sqrt{2}}.$$

Note that the coefficients g_0 and g_1 form the 2-period simple moving average with unit energy, and the coefficients h_0 and h_1 form the 2-point moving difference with unit energy. Another example is the order 2 Daubechies wavelet, which has the coefficients with $M = 4$ [Daubechies (1992); Gençay *et al.* (2001)]

$$h_0 = \frac{1-\sqrt{3}}{4\sqrt{2}}, \quad h_1 = \frac{-3+\sqrt{3}}{4\sqrt{2}}, \quad h_2 = \frac{3+\sqrt{3}}{4\sqrt{2}}, \quad \text{and} \quad h_3 = \frac{-1-\sqrt{3}}{4\sqrt{2}},$$

and

$$g_0 = \frac{1+\sqrt{3}}{4\sqrt{2}}, \quad g_1 = \frac{3+\sqrt{3}}{4\sqrt{2}}, \quad g_2 = \frac{3-\sqrt{3}}{4\sqrt{2}}, \quad \text{and} \quad g_3 = \frac{1-\sqrt{3}}{4\sqrt{2}}.$$

Given the input data as a vector, the DWT is expressed by the matrix-vector multiplication. For instance, the Haar wavelet filter for a 2-point input signal has the following matrix operator

$$H_2 = \frac{1}{\sqrt{2}} \begin{bmatrix} 1 & -1 \\ 1 & 1 \end{bmatrix},$$

where the first row of H_2 corresponds to the mother wavetlet (high-pass filtering) and the second row of H_2 corresponds to the father wavelet (low-pass filtering). Suppose we consider a simple 2-point signal $\mathbf{x} = [1, 2]^T$. Then, the output from the Haar wavelet filter is

$$\mathbf{y} = H_2\mathbf{x} = \frac{1}{\sqrt{2}} \begin{bmatrix} 1 & -1 \\ 1 & 1 \end{bmatrix} \begin{bmatrix} 1 \\ 2 \end{bmatrix} = \begin{bmatrix} -\frac{1}{\sqrt{2}} \\ \frac{3}{\sqrt{2}} \end{bmatrix}.$$

Note that H_2 is an orthogonal matrix, i.e., $H_2 H_2^T = I_2$, where I_2 is a 2×2 identity matrix. Then the signal $\mathbf{x}$ can be reconstructed by the inverse operation

$$\mathbf{x} = H_2^T \mathbf{y} = \frac{1}{\sqrt{2}} \begin{bmatrix} 1 & 1 \\ -1 & 1 \end{bmatrix} \begin{bmatrix} -\frac{1}{\sqrt{2}} \\ \frac{3}{\sqrt{2}} \end{bmatrix} = \begin{bmatrix} 1 \\ 2 \end{bmatrix}.$$

If we look at it the other way, the first column of H_2^T (or equivalently the first row of H_2) with the coordinate $-\frac{1}{\sqrt{2}}$ plus the second column of H_2^T (or equivalently the second row of H_2) with the coordinate $\frac{3}{\sqrt{2}}$ gives the original input signal

$$\mathbf{x} = \begin{bmatrix} 1 \\ 2 \end{bmatrix} = -\frac{1}{\sqrt{2}} \begin{bmatrix} \frac{1}{\sqrt{2}} \\ -\frac{1}{\sqrt{2}} \end{bmatrix} + \frac{3}{\sqrt{2}} \begin{bmatrix} \frac{1}{\sqrt{2}} \\ \frac{1}{\sqrt{2}} \end{bmatrix} \equiv \mathbf{d}_1 + \mathbf{r}.$$

Here the vector $\mathbf{d}_1 = [-\frac{1}{2}, \frac{1}{2}]^T$ stands for the deviation of $\mathbf{x} = [1, 2]^T$ away from its mean vector $\mathbf{r} = [\frac{3}{2}, \frac{3}{2}]^T$.

To achieve different level of scales, we manipulate the dilation and translation parameters. For instance, the Haar wavelet filter for a 8-point input signal has the following matrix operator as its DWT:

$$H_8 = \frac{1}{2\sqrt{2}} \begin{bmatrix} 2 & -2 & 0 & 0 & 0 & 0 & 0 & 0 \\ 0 & 0 & 2 & -2 & 0 & 0 & 0 & 0 \\ 0 & 0 & 0 & 0 & 2 & -2 & 0 & 0 \\ 0 & 0 & 0 & 0 & 0 & 0 & 2 & -2 \\ \sqrt{2} & \sqrt{2} & -\sqrt{2} & -\sqrt{2} & 0 & 0 & 0 & 0 \\ 0 & 0 & 0 & 0 & \sqrt{2} & \sqrt{2} & -\sqrt{2} & -\sqrt{2} \\ 1 & 1 & 1 & 1 & -1 & -1 & -1 & -1 \\ 1 & 1 & 1 & 1 & 1 & 1 & 1 & 1 \end{bmatrix}.$$

Here, the entries in H_8 are scaled such that H_8 is an orthogonal matrix, i.e., $H_8 H_8^T = I_8$, where I_8 is an identity matrix of size 8×8. It is still observable that the first six rows of H_8 correspond to the daughter wavelets, the seventh row correspond to the mother wavelet, and the last row correspond to the father wavelet. Given an 8-point signal $\mathbf{x} = [1, 2, 3, 4, 4, 3, 2, 1]^T$, the output is found to be

$$\mathbf{y} = H_8\mathbf{x}$$

$$= \frac{1}{2\sqrt{2}} \begin{bmatrix} 2 & -2 & 0 & 0 & 0 & 0 & 0 & 0 \\ 0 & 0 & 2 & -2 & 0 & 0 & 0 & 0 \\ 0 & 0 & 0 & 0 & 2 & -2 & 0 & 0 \\ 0 & 0 & 0 & 0 & 0 & 0 & 2 & -2 \\ \sqrt{2} & \sqrt{2} & -\sqrt{2} & -\sqrt{2} & 0 & 0 & 0 & 0 \\ 0 & 0 & 0 & 0 & \sqrt{2} & \sqrt{2} & -\sqrt{2} & -\sqrt{2} \\ 1 & 1 & 1 & 1 & -1 & -1 & -1 & -1 \\ 1 & 1 & 1 & 1 & 1 & 1 & 1 & 1 \end{bmatrix} \begin{bmatrix} 1 \\ 2 \\ 3 \\ 4 \\ 4 \\ 3 \\ 2 \\ 1 \end{bmatrix} = \begin{bmatrix} -\frac{1}{\sqrt{2}} \\ -\frac{1}{\sqrt{2}} \\ \frac{1}{\sqrt{2}} \\ \frac{1}{\sqrt{2}} \\ -2 \\ 2 \\ 0 \\ \frac{10}{\sqrt{2}} \end{bmatrix}.$$

Then, the signal $\mathbf{x}$ can be reconstructed by the inverse operation

$$\mathbf{x} = H_8^T\mathbf{y}$$

$$= \frac{1}{2\sqrt{2}} \begin{bmatrix} 2 & 0 & 0 & 0 & \sqrt{2} & 0 & 1 & 1 \\ -2 & 0 & 0 & 0 & \sqrt{2} & 0 & 1 & 1 \\ 0 & 2 & 0 & 0 & -\sqrt{2} & 0 & 1 & 1 \\ 0 & -2 & 0 & 0 & -\sqrt{2} & 0 & 1 & 1 \\ 0 & 0 & 2 & 0 & 0 & \sqrt{2} & -1 & 1 \\ 0 & 0 & -2 & 0 & 0 & \sqrt{2} & -1 & 1 \\ 0 & 0 & 0 & 2 & 0 & -\sqrt{2} & -1 & 1 \\ 0 & 0 & 0 & -2 & 0 & -\sqrt{2} & -1 & 1 \end{bmatrix} \begin{bmatrix} -\frac{1}{\sqrt{2}} \\ -\frac{1}{\sqrt{2}} \\ \frac{1}{\sqrt{2}} \\ \frac{1}{\sqrt{2}} \\ -2 \\ 2 \\ 0 \\ \frac{10}{\sqrt{2}} \end{bmatrix} = \begin{bmatrix} 1 \\ 2 \\ 3 \\ 4 \\ 4 \\ 3 \\ 2 \\ 1 \end{bmatrix}.$$

Similarly, through the reconstruction of $\mathbf{x}$, we see that it is formed by the linear combinations of columns of H_8^T (or rows of H_8) multiplied by the corresponding entries (coordinates) from the output $\mathbf{y}$. Consequently, we obtain the following decomposition for $\mathbf{x}$:

$$\mathbf{x} = \sum_{k=1}^{3} \mathbf{d}_k + \mathbf{r},$$

where

$$\mathbf{d}_1 = \begin{bmatrix} -\frac{1}{2} \\ \frac{1}{2} \\ -\frac{1}{2} \\ \frac{1}{2} \\ \frac{1}{2} \\ -\frac{1}{2} \\ \frac{1}{2} \\ -\frac{1}{2} \end{bmatrix} = -\frac{1}{\sqrt{2}} \begin{bmatrix} \frac{1}{\sqrt{2}} \\ -\frac{1}{\sqrt{2}} \\ 0 \\ 0 \\ 0 \\ 0 \\ 0 \\ 0 \end{bmatrix} - \frac{1}{\sqrt{2}} \begin{bmatrix} 0 \\ 0 \\ \frac{1}{\sqrt{2}} \\ -\frac{1}{\sqrt{2}} \\ 0 \\ 0 \\ 0 \\ 0 \end{bmatrix} + \frac{1}{\sqrt{2}} \begin{bmatrix} 0 \\ 0 \\ 0 \\ 0 \\ \frac{1}{\sqrt{2}} \\ -\frac{1}{\sqrt{2}} \\ 0 \\ 0 \end{bmatrix} + \frac{1}{\sqrt{2}} \begin{bmatrix} 0 \\ 0 \\ 0 \\ 0 \\ 0 \\ 0 \\ \frac{1}{\sqrt{2}} \\ -\frac{1}{\sqrt{2}} \end{bmatrix},$$

$$\mathbf{d}_2 = \begin{bmatrix} -1 \\ -1 \\ 1 \\ 1 \\ 1 \\ 1 \\ -1 \\ -1 \end{bmatrix} = -2 \begin{bmatrix} \frac{1}{2} \\ \frac{1}{2} \\ -\frac{1}{2} \\ -\frac{1}{2} \\ 0 \\ 0 \\ 0 \\ 0 \end{bmatrix} + 2 \begin{bmatrix} 0 \\ 0 \\ 0 \\ 0 \\ \frac{1}{2} \\ \frac{1}{2} \\ -\frac{1}{2} \\ -\frac{1}{2} \end{bmatrix},$$

$$\mathbf{d}_3 = \begin{bmatrix} 0 \\ 0 \\ 0 \\ 0 \\ 0 \\ 0 \\ 0 \\ 0 \end{bmatrix} = 0 \begin{bmatrix} \frac{1}{2\sqrt{2}} \\ \frac{1}{2\sqrt{2}} \\ \frac{1}{2\sqrt{2}} \\ \frac{1}{2\sqrt{2}} \\ -\frac{1}{2\sqrt{2}} \\ -\frac{1}{2\sqrt{2}} \\ -\frac{1}{2\sqrt{2}} \\ -\frac{1}{2\sqrt{2}} \end{bmatrix} \quad \text{and} \quad \mathbf{r} = \begin{bmatrix} \frac{5}{2} \\ \frac{5}{2} \\ \frac{5}{2} \\ \frac{5}{2} \\ \frac{5}{2} \\ \frac{5}{2} \\ \frac{5}{2} \\ \frac{5}{2} \end{bmatrix} = \frac{10}{\sqrt{2}} \begin{bmatrix} \frac{1}{2\sqrt{2}} \\ \frac{1}{2\sqrt{2}} \\ \frac{1}{2\sqrt{2}} \\ \frac{1}{2\sqrt{2}} \\ \frac{1}{2\sqrt{2}} \\ \frac{1}{2\sqrt{2}} \\ \frac{1}{2\sqrt{2}} \\ \frac{1}{2\sqrt{2}} \end{bmatrix}.$$

Again, the vector $\mathbf{r}$ gives the mean of the input signal $\mathbf{x}$, while $\mathbf{d}_1$, $\mathbf{d}_2$, and $\mathbf{d}_3$ give different levels of fluctuation away from $\mathbf{r}$; see Figure 8.1 for a graphical illustration of the decomposition.

Haar wavelet filter is not the only tool used in wavelet decomposition. More generally, one uses various wavelet filters through scaling and dilation

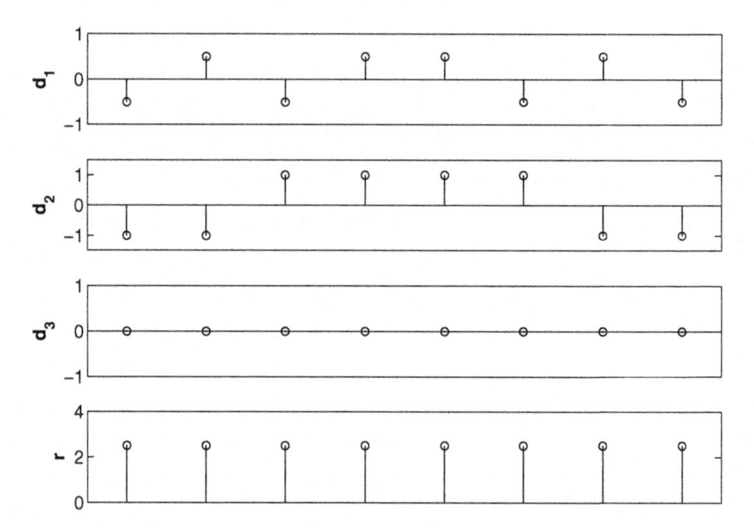

Figure 8.1: Multilevel wavelet decomposition (with $m = 3$) with the Haar wavelet on $\mathbf{x} = [1, 2, 3, 4, 4, 3, 2, 1]^T$

to decompose a given input signal into components of multilevel frequency. Such technique is known as the wavelet multiresolution analysis (MRA). In this context, we do not go into the details of MRA because it is a well-known subject which can be found in existing literature. Readers are referred to Daubechies (1992); Mallet (2008); Gençay *et al.* (2001) for more topics about wavelet decomposition and also its fast algorithms. Given a set of observable data $\mathbf{x}$, the multilevel wavelet decomposition (with m levels) is then

$$\mathbf{x} = \sum_{k=1}^{m} \mathbf{d}_k + \mathbf{r}, \tag{8.3}$$

where each $\mathbf{d}_k$ for $k = 1, 2, \ldots, m$ is called the kth level wavelet detail extracted by the DWT, and the residual term $\mathbf{r}$ is the mean or trend of the input signal $\mathbf{x}$. Figure 8.2 shows the wavelet decomposition using the order 2 Daubechies wavelet for the Hang Seng Index from 2007 to 2011. The decomposition level is chosen as $m = 3$. The first row is the first wavelet detail $\mathbf{d}_1$ with the highest frequency, in other words, the dynamics and movements of the Hang Seng Index. The second and third rows are the second and third wavelet details $\mathbf{d}_2$ and $\mathbf{d}_3$ with relatively lower frequencies. The last row is the trend $\mathbf{r}$, which is the smoothed version of the original data.

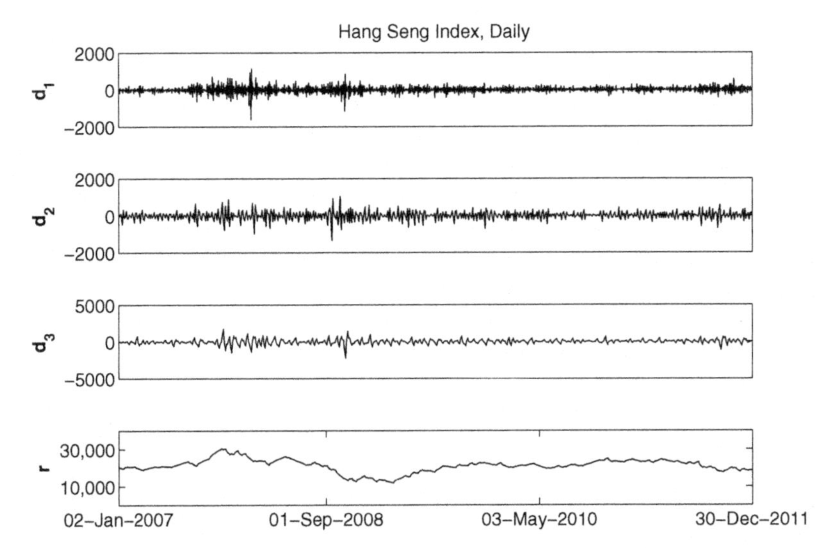

Figure 8.2: Multilevel wavelet decomposition (with $m = 3$) with the order 2 Daubechies wavelet on Hang Seng Index from 2007 to 2011

In general, if we combine the residual term $\mathbf{r}$ with the last wavelet detail $\mathbf{d}_m$, a slow-varying component of the original signal is produced. When we add more wavelet details to the residual term $\mathbf{r}$, we can gradually reconstruct the original signal. By changing how many wavelet details are being summed, we derive different smoothed versions of the original input with preferred frequency range. This idea will be adopted to design a wavelet-based trading method in Section 8.3.

8.3 Trading Method Based on Wavelets

Our subject of interest is an n-length observation of the index closing prices, denoted by $(x_1, x_2, \ldots, x_n)$, where x_j is the index closing price on day j for $j = 1, 2, \ldots, n$. We measure the profitability of the wavelet-based indicator during this n-length period. In other words, we first need to classify each day as either a buy, sell, or neutral day. To achieve this, for each day $j = 1, 2, \ldots, n$, we extend the subject period with some past data:

$$\mathbf{x}_j = (x_{-n_0}, \ldots, x_{-1}, x_0, x_1, x_2, \ldots, x_j),$$

where $n_0 + 1$ is the number of trading days in the year before the n-length observation period (here x_1 is chosen as the first trading day in a certain year). Then, we perform the multilevel wavelet decomposition with level m

for $\mathbf{x}_j$ to obtain

$$\mathbf{x}_j = \sum_{k=1}^{m} \mathbf{d}_k^{(j)} + \mathbf{r}^{(j)}$$

for day j. Here $\mathbf{d}_k^{(j)}$ and $\mathbf{r}^{(j)}$ are vectors of the same length as the input $\mathbf{x}_j$:

$$\mathbf{d}_k^{(j)} = (d_{k,-n_0}^{(j)}, \ldots, d_{k,0}^{(j)}, d_{k,1}^{(j)}, \ldots, d_{k,j}^{(j)})$$

and

$$\mathbf{r}^{(j)} = (r_{-n_0}^{(j)}, \ldots, r_0^{(j)}, r_1^{(j)}, \ldots, r_j^{(j)}).$$

Note that the decomposition differs for each day j and the notations (j) are used to show that the decomposition depends on day j. The trading indicator designed next is based on such a multilevel wavelet decomposition.

Similar to the moving average crossover rule, we will make use of two indicating lines to generate buying and selling signals. These two lines are also referred to as the short line and the long line. The short line represents the component with a shorter cycle, while the long line represents the component with a longer cycle. Then, the difference between these two lines produces a band-pass filter, or momentum indicator, which leads to a series of market entries and exits. To design a wavelet-based technical indicator, we manipulate the number of wavelet details added on the mean vector. By changing this number, we derive approximations to the original input data. To replicate the moving average crossover rule, we use two different partial sums of the multilevel wavelet decomposition to construct the short line and the long line.

For day j, we define the following value to construct the short line

$$MRMA_{j_1,k_1,j} = \sum_{k=k_1}^{m} d_{k,j-j_1}^{(j)} + r_{j-j_1}^{(j)}, \quad j = 1, 2, \ldots, n.$$

Here, MRMA is the short term for the combination of multiresolution and moving average. As j varies from 1 to n over the n-length observation period, all the values $MRMA_{j_1,k_1,j}$ will then form the short line MRMA_{j_1,k_1}. To generate the short line, we need to sum up most of the wavelet details. Therefore, the parameter k_1 is chosen as an integer close to or equal to 1, to obtain a component of shorter cycle. The second parameter j_1 is used to shift the short line backwards in time. We use an older value at time $j - j_1$ instead of j on the short line such that crossovers with the long line are created. Note that when $k_1 = 1$ and $j_1 = 0$ the short line is reduced to the original data. Similarly, we define the following value on the long line for each day j:

$$MRMA_{j_2,k_2,j} = \sum_{k=k_2}^{m} d_{k,j-j_2}^{(j)} + r_{j-j_2}^{(j)}, \quad j = 1, 2, \ldots, n,$$

where we have added fewer wavelet details in the summation with $k_2 > k_1$ to get a component with longer cycle, and correspondingly we select $j_2 > j_1$ to make the long line lag further in time to create crossovers with the short line. Then, over the n-length observation period, all the values $MRMA_{j_2,k_2,j}$, $j = 1, 2, \ldots, n$ form the long line MRMA_{j_2,k_2}.

With the short line and the long line defined on the n-length observation period, we now assume that a day j belongs to the buy period if

$$MRMA_{j_1,k_1,j} > MRMA_{j_2,k_2,j} \times \left(1 + \frac{p}{100}\right)$$

and a day j belongs to the sell period if

$$MRMA_{j_1,k_1,j} < MRMA_{j_2,k_2,j} \times \left(1 - \frac{p}{100}\right).$$

Here a $p\%$ band is incorporated into the long line to reduce frequent trading positions. We measure the profitability of this wavelet-based indicator in Section 8.4.

8.4 Evaluation Results

We consider the daily Hang Seng Index and China Shanghai Composite Index exported from Yahoo! Finance for the period from January 2007 to December 2011. We measure the performance of the wavelet-based indicator by studying test statistics of profits generated by its buy and sell signals. We use the symbol $(k_1, j_1, k_2, j_2, p\%)$ to denote the wavelet-based trading strategy, where the parameters k_1 and j_1 govern the short line, k_2 and j_2 the long line, and p the band. Figure 8.3 gives an example of short and long lines derived from the multilevel wavelet decomposition of Hang Seng Index in the corresponding period. Note that once again when $k_1 = 1$ and $j_1 = 0$, the short line is merely the input data. Lastly, we perform all the wavelet decompositions with $m = 5$ levels in the following numerical examples. The tested wavelet filters are Haar wavelet, order 2 and order 4 Daubechies wavelets [Daubechies (1992)].

Tables 8.1 and 8.2 report the mean daily return $\hat{\mu}_{\mathrm{b}}$ from the long period with the corresponding test statistic T_{b}, the mean daily return $\hat{\mu}_{\mathrm{s}}$ from the

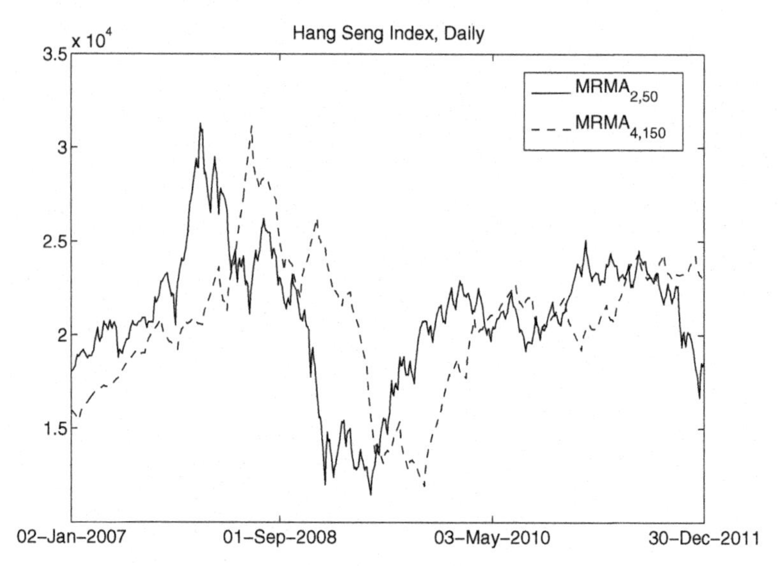

Figure 8.3: The short line and the long line generated by the wavelet-based trading rule with order 2 Daubechies wavelet for the Hang Seng Index from 2007 to 2011

short period with the corresponding test statistic T_{s}, the mean return difference $\hat{\mu}_{\text{b}} - \hat{\mu}_{\text{s}}$ with the corresponding test statistic T_{bs}, and the break-even costs C; see Section 7.2 for more details. In addition, the number of buy days and sell days N_{b} and N_{s} generated by the wavelet-based trading strategy are reported in the tables.

In Table 8.1, we observe that the mean daily returns $\hat{\mu}_{\text{b}}$'s for the long strategy are all positive, ranging from 0.052 to 0.080%. Overall, the wavelet-based indicators generate profits similar to the moving average family in the Hong Kong stock market; see Table 7.1. Among the tested wavelet-based indicators, the rule $(3, 50, 5, 150, 1\%)$ with the order 2 Daubechies wavelet gives the best mean daily return of 0.080%, or 20.16% annually. Table 8.1 also displays the performance of the short strategy generated by the wavelet-based trading rules for Hang Seng Index. Their mean returns $\hat{\mu}_{\text{s}}$'s are all negative.

In Table 8.2, another set of wavelet-based trading rules is tested on the China Shanghai Composite Index, from the period of 2007 to 2011. Compared with Table 8.1, most of the results in Table 8.2 have higher daily mean returns and significant levels. This is similar to how Table 7.2 dominates Table 7.1 because of market efficiencies in two places. Among the tested wavelet-based indicators, the rule $(1, 0, 3, 50, 0\%)$ with order 2 Daubechies wavelet gives the best mean daily return of 0.169%, or 42.59% annually.

Table 8.1: Standard test results and break-even costs for Hang Seng Index from 2007 to 2011, with wavelet-based trading rules

Rule	N_b	N_s	$\hat{\mu}_b$ (T_b)	$\hat{\mu}_s$ (T_s)	$\hat{\mu}_b - \hat{\mu}_s$ (T_{bs})	C
			'db1' (Haar)			
(1, 0, 5, 150, 0%)	741	492	0.00060 (0.80)	−0.00110 (−0.79)	0.00170 (1.30)	0.0411
(1, 0, 5, 150, 1%)	716	475	0.00057 (0.76)	−0.00108 (−0.76)	0.00165 (1.23)	0.0440
(2, 15, 5, 150, 0%)	732	501	0.00058 (0.77)	−0.00104 (−0.75)	0.00162 (1.26)	0.0590
(2, 15, 5, 150, 1%)	714	487	0.00048 (0.65)	−0.00115 (−0.83)	0.00164 (1.24)	0.0533
(3, 50, 5, 150, 0%)	699	534	0.00063 (0.81)	−0.00100 (−0.76)	0.00163 (1.31)	0.0609
(3, 50, 5, 150, 1%)	679	506	0.00068 (0.87)	−0.00116 (−0.85)	0.00184 (1.41)	0.0748
			'db2'			
(1, 0, 5, 150, 0%)	737	496	0.00053 (0.72)	−0.00099 (−0.71)	0.00152 (1.17)	0.0551
(1, 0, 5, 150, 1%)	726	488	0.00069 (0.90)	−0.00099 (−0.70)	0.00167 (1.28)	0.0577
(2, 15, 5, 150, 0%)	727	506	0.00058 (0.76)	−0.00102 (−0.75)	0.00160 (1.25)	0.0668
(2, 15, 5, 150, 1%)	712	497	0.00071 (0.91)	−0.00097 (−0.70)	0.00168 (1.29)	0.0659
(3, 50, 5, 150, 0%)	698	535	0.00073 (0.93)	−0.00114 (−0.87)	0.00187 (1.51)	0.0701
(3, 50, 5, 150, 1%)	674	504	0.00080 (1.00)	−0.00118 (−0.87)	0.00199 (1.53)	0.0669
			'db4'			
(1, 0, 5, 150, 0%)	739	494	0.00059 (0.78)	−0.00107 (−0.77)	0.00166 (1.28)	0.0803
(1, 0, 5, 150, 1%)	728	485	0.00059 (0.78)	−0.00104 (−0.74)	0.00163 (1.24)	0.0666
(2, 15, 5, 150, 0%)	730	503	0.00058 (0.76)	−0.00103 (−0.75)	0.00160 (1.25)	0.0521
(2, 15, 5, 150, 1%)	718	487	0.00052 (0.69)	−0.00108 (−0.77)	0.00160 (1.21)	0.0599
(3, 50, 5, 150, 0%)	706	527	0.00067 (0.87)	−0.00108 (−0.81)	0.00175 (1.40)	0.0653
(3, 50, 5, 150, 1%)	678	509	0.00066 (0.84)	−0.00105 (−0.77)	0.00170 (1.32)	0.0575

Table 8.2: Standard test results and break-even costs for China Shanghai Composite Index from 2007 to 2011, with wavelet-based trading rules

Rule	N_b	N_s	$\hat{\mu}_b$ (T_b)	$\hat{\mu}_s$ (T_s)	$\hat{\mu}_b - \hat{\mu}_s$ (T_{bs})	C
			'db1' (Haar)			
(1, 0, 3, 50, 0%)	586	629	0.00157 (1.77)	−0.00180 (−1.59)	0.00337 (2.92)	0.0684
(1, 0, 3, 50, 1%)	563	617	0.00153 (1.73)	−0.00200 (−1.77)	0.00354 (3.03)	0.0778
(1, 0, 4, 100, 0%)	649	566	0.00112 (1.38)	−0.00166 (−1.38)	0.00278 (2.37)	0.0596
(1, 0, 4, 100, 1%)	621	546	0.00112 (1.34)	−0.00178 (−1.46)	0.00290 (2.40)	0.0557
(2, 15, 4, 100, 0%)	655	560	0.00121 (1.45)	−0.00179 (−1.52)	0.00300 (2.56)	0.0598
(2, 15, 4, 100, 1%)	624	540	0.00141 (1.62)	−0.00172 (−1.42)	0.00313 (2.59)	0.0646
			'db2'			
(1, 0, 3, 50, 0%)	580	635	0.00169 (1.90)	−0.00187 (−1.66)	0.00356 (3.09)	0.0775
(1, 0, 3, 50, 1%)	559	617	0.00149 (1.68)	−0.00177 (−1.54)	0.00326 (2.78)	0.0621
(1, 0, 4, 100, 0%)	641	574	0.00125 (1.50)	−0.00176 (−1.49)	0.00301 (2.58)	0.0755
(1, 0, 4, 100, 1%)	619	561	0.00130 (1.53)	−0.00183 (−1.53)	0.00314 (2.63)	0.0875
(2, 15, 4, 100, 0%)	651	564	0.00131 (1.55)	−0.00189 (−1.62)	0.00319 (2.74)	0.0639
(2, 15, 4, 100, 1%)	630	541	0.00129 (1.50)	−0.00202 (−1.70)	0.00331 (2.74)	0.1002
			'db4'			
(1, 0, 3, 50, 0%)	582	633	0.00163 (1.85)	−0.00183 (−1.62)	0.00347 (3.01)	0.0812
(1, 0, 3, 50, 1%)	555	615	0.00150 (1.69)	−0.00195 (−1.72)	0.00346 (2.94)	0.0754
(1, 0, 4, 100, 0%)	639	576	0.00132 (1.58)	−0.00183 (−1.55)	0.00315 (2.70)	0.0949
(1, 0, 4, 100, 1%)	619	558	0.00125 (1.48)	−0.00180 (−1.50)	0.00305 (2.55)	0.0742
(2, 15, 4, 100, 0%)	653	562	0.00128 (1.52)	−0.00186 (−1.59)	0.00314 (2.68)	0.1044
(2, 15, 4, 100, 1%)	625	542	0.00121 (1.42)	−0.00189 (−1.58)	0.00310 (2.57)	0.0991

For the short strategy generated by the wavelet-based trading rules for the Chinese stock market, the mean returns μ_s's are also all negative.

8.5 Summary

In this chapter, the wavelets are introduced and applied to design a technical indicator. The trading strategy is to first decompose the signal by multiresolution analysis. Then, we monitor the crossovers of two different components reconstructed with different level of frequency, i.e., different number of wavelet details being added. The idea, which resembles the moving average crossover rule, comes from the momentum indicators in Chapter 6. Numerical results show that the wavelet-based trading rules can perform decently well in the Hong Kong and Chinese stock markets over the last five years.

Chapter 9

Empirical Mode Decomposition and Technical Analysis

It is well known that nonlinearity and non-stationarity are commonly observed in financial time series. Classic Fourier analysis assumes the time series to be linear and stationary. On the other hand, the more advanced wavelet analysis is proposed for non-stationary time series, but the linearity issue remains. Data analysis methods for nonlinear and non-stationary processes are relatively scarce.

The Hilbert–Huang transform (HHT), proposed by Norden E. Huang in 1998 [Huang *et al.* (1998)], is an advanced analysis tool for nonlinear and non-stationary data. The Hilbert–Huang transform is composed of two parts: empirical mode decomposition (EMD) and Hilbert spectrum analysis (HSA). The EMD is a data-adaptive algorithm which decomposes a complicated real-world signal into a small number of components known as the intrinsic mode functions (IMFs). The IMFs have well-behaved instantaneous frequency, which unveils more signal behaviors because it varies in time. In this context, we focus on applying the EMD algorithm in the technical analysis field, and consider how to derive a trading strategy based on the decomposition.

9.1 Instantaneous Frequency

Classic frequency is defined as the reciprocal of the periodicity. For instance, the cosine function has a clear periodic behavior because it always takes 2π to finish a cycle. Then the classic frequency can be easily determined as the reciprocal $\frac{1}{2\pi}$. Equivalently the angular frequency is 1, which is the classic frequency $\frac{1}{2\pi}$ multiplied by 2π. However, in real-time data analysis, periodicity is buried under the noise and it is difficult to define the classic frequency. Moreover, the classic frequency does not change over time, which is unrealistic in actual applications.

Practitioners then resort to the instantaneous frequency, or time-varying frequency. In the following context, we use the notation t to denote a continuous time variable. The notation j which is used to denote a discrete time variable in the previous context will be adopted later on. Given a real-valued signal $x(t)$, we first find its Hilbert transform $y(t)$ by:

$$y(t) = \frac{1}{\pi}\text{PV}\int_{-\infty}^{\infty} \frac{x(\tau)}{t-\tau}d\tau,$$

where PV represents the Cauchy principal value. With the Hilbert transform $y(t)$, we form an analytic signal by

$$z(t) = x(t) + iy(t) = a(t)\exp(i\theta(t)),$$

where

$$a(t) = \sqrt{x^2(t) + y^2(t)} \quad \text{and} \quad \theta(t) = \arctan\left(\frac{y(t)}{x(t)}\right).$$

Here $a(t)$ is the instantaneous amplitude, $\theta(t)$ is the instantaneous phase, and they both depend on time. The instantaneous frequency is then defined by

$$\omega(t) = \frac{d\theta(t)}{dt}.$$

This definition of instantaneous frequency works well with a sine or cosine function. Suppose we have the real-valued signal $x(t) = \cos(t)$. Its Hilbert transform is exactly given by $y(t) = \sin(t)$. Therefore, the analytic signal would be $z(t) = \exp(it)$ and the instantaneous frequency turns out to be a constant independent of time:

$$\omega(t) = \frac{dt}{dt} = 1.$$

In this case, the instantaneous frequency of $\cos(t)$ agrees to its angular frequency.

However, this definition of instantaneous frequency does not always work well, especially when the given real-valued signal has a drift [Huang *et al.* (1998)]. For example, suppose we consider $x(t) = \sin(t)+b$, where b is a constant. Intuitively, the frequency of $x(t)$ should stay the same, regardless of the value of the shifting parameter b. According to Huang *et al.* (1998), nevertheless, if we directly apply the Hilbert transform for $x(t)$, its instantaneous frequency calculated by the above-mentioned manner might be different when we vary b. At times, negative frequency values could appear for a certain value of b. Therefore, Huang *et al.* (1998) propose to process the signal beforehand and decompose it into several components that have no drifting movement. These components will be well-behaved after applying Hilbert transform, and admit to more standard instantaneous frequencies.

9.2 Empirical Mode Decomposition

Recall that the classic frequency works well for functions like sine and cosine because their periodicities are well-defined with observable cycles. Nevertheless, it is difficult to define the periodicity of a nonlinear and nonstationary signal over a time horizon. Instead of using cycles to define a period, alternatively we can monitor the zero-crossings and extrema of the signal. Suppose we are given a real-valued signal $x(t) = \sin(t)$. When we study the signal $x(t)$ on any segment of the horizontal axis t, it can be observed that the number of zero-crossings is equal to or differed at most by one from the number of extrema. Moreover, the signal is "symmetric" about the horizontal axis, i.e., the upper and lower envelops that encompass $x(t) = \sin(t)$ have a mean value of zero. With such properties, a new class of functions known as the IMFs is introduced in Huang *et al.* (1998). IMFs are defined using the following two properties:

(1) the number of extrema and the number of zero-crossings are either equal or differ at most by one, and
(2) the mean value of the upper envelope (connected by local maxima) and the lower envelope (connected by local minima) is zero.

According to the definition, sine and cosine functions are already IMFs. IMFs in general resemble sinusoidal functions in shape, but there are two visible major differences. Firstly, sinusoidal functions have a fixed amplitude no matter how time changes, and secondly, they take the same amount of time to finish a traditional cycle (traveling from a zero-crossing to a peak, the next zero-crossing, a trough, then back to a zero-crossing). In Figure 9.1, an IMF extracted from Hang Seng Index (performed by an algorithm which will be discussed later) is illustrated. We observe that such an IMF does not have a fixed amplitude when time advances. Moreover, the time an IMF takes to finish a traditional cycle is different, and this gives birth to

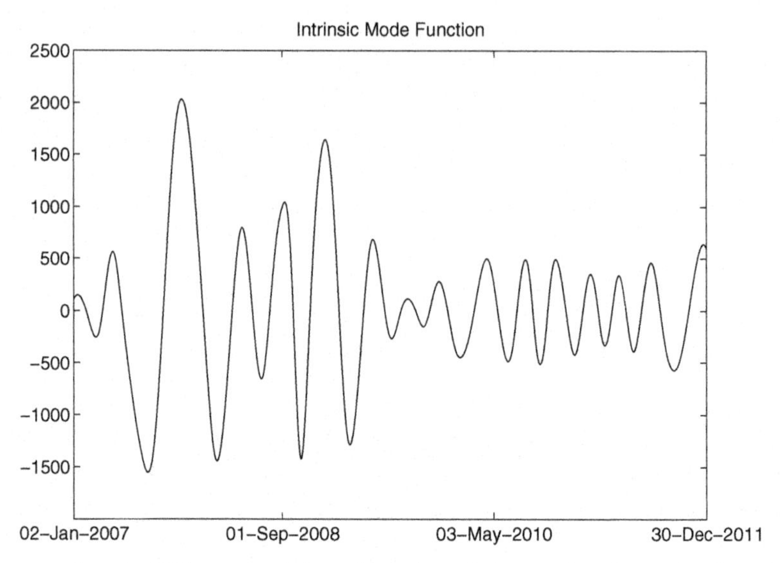

Figure 9.1: An IMF extracted from Hang Seng Index during the period from 2007 to 2011

variable frequency. Thus, IMFs are deemed to be the generalization of sine and cosine functions, the basis functions in Fourier analysis, and therefore have wider properties.

In Fourier analysis, we decompose a signal into sinusoidal functions, which have stable characteristics in the frequency domain. For nonlinear and non-stationary signals, the aim is to decompose a signal into a finite (often small) number of IMFs instead of sinusoidal functions. However, IMFs are not defined mathematically, but literally via two properties that they have to hold. Therefore, we do not have a mathematical way like Fourier transform to decompose a signal into IMFs. EMD, the empirical mode decomposition, serves a role similar to the Fourier transform in Fourier analysis, but instead it is an algorithm that extracts IMFs step by step from the original signal. Assume now we consider a discrete input signal $\mathbf{x} = (x_1, x_2, \ldots, x_n)$ of length n. The procedures for the EMD algorithm are introduced in the following [Huang *et al.* (1998)].

(1) Identify all the maxima and minima of $\mathbf{x}$;
(2) Connect all maxima and minima by cubic spline lines, thus form the upper envelope and the lower envelope respectively;
(3) Calculate the local mean, denoted by $(a_1, a_2, \ldots, a_n)$, from the upper and lower envelopes;

(4) Extract the detail $\mathbf{d} = (d_1, d_2, \ldots, d_n)$ where $d_j = x_j - a_j$ for $j = 1, 2, \ldots, n$;
(5) Check if $\mathbf{d}$ satisfies the definition of an IMF; If not, repeat steps (1)–(4) (with $\mathbf{d}$ being the new $\mathbf{x}$) till the detail meets the conditions of an IMF;
(6) The detail which meets the two IMF conditions is denoted as the first IMF $\mathbf{c}_1 = (c_{1,1}, c_{1,2}, \ldots, c_{1,n})$. Iterate the routines on the residual $\mathbf{x} - \mathbf{c}_1$ to obtain the second IMF (which will be denoted as $\mathbf{c}_2 = (c_{2,1}, c_{2,2}, \ldots, c_{2,n})$). The kth extracted IMF will be denoted as $\mathbf{c}_k = (c_{k,1}, c_{k,2}, \ldots, c_{k,n})$.
(7) End the operation when the residue cannot be decomposed anymore.

In the above routine, step (5) is also known as the sifting process. It is used to alleviate the overshoots and undershoots during the process of obtaining an IMF [Huang *et al.* (1998)]. Consequently, the original time series $\mathbf{x}$ is decomposed as the sum of m IMFs $\mathbf{c}_k$ together with the residual $\mathbf{r}$ that cannot be decomposed anymore:

$$\mathbf{x} = \sum_{k=1}^{m} \mathbf{c}_k + \mathbf{r}. \tag{9.1}$$

The number of IMFs extracted from a signal, m, is approximately of order $\log_2(n)$, where n is the length of the input data. According to the EMD algorithm, the first IMF $\mathbf{c}_1$ should represent the highest-frequency component because the local mean with low frequency has been iteratively removed from the original data. As the procedures continue, the subsequent IMFs have lower and lower frequency ranges. For the residue $\mathbf{r}$, it is not an IMF, but represents the mean trend of the whole data. In some special cases, the residual can even be a straight line.

One can observe the similarity between the EMD (9.1) and the multilevel wavelet decomposition (8.3) in Section 8.2. The wavelet details $\mathbf{d}_k$'s in (8.3) represent components of the original signal with different frequency (lower frequency with a larger k). On the other hand, the IMFs ($\mathbf{c}_k$'s) also play a similar role in (9.1). In both decompositions, there is an $\mathbf{r}$ which shows the mean or the trend of the input signal. However, the EMD depends on the input data, while the multilevel wavelet decomposition depends on the chosen wavelet filter.

The EMD is easy to implement and it is a data-adaptive method. More importantly, all the IMFs admit to better-behaved Hilbert transform and instantaneous frequency [Huang *et al.* (1998)]. Since the instantaneous frequency on each IMF is well-defined, we can extract interesting features such as seasonality or business cycle. The IMFs can also be regarded as the filtered output of the original signal. For instance, we consider

$$\sum_{k=k_0}^{m} \mathbf{c}_k + \mathbf{r},$$

where k_0 is a number close to m. Since we are summing up the last few IMFs and the residual, the summation can be seen as an output by a low-pass filtering algorithm. To generalize this idea, one can drop out the unwanted IMFs in the full summation to obtain an output with a certain range of frequencies. Note that the EMD is an *a posterior* data processing method and there is no way to write out the explicit form for such filtering operation, unless the EMD is replaced by some sort of alternative iterative method [Lin *et al.* (2009)]. In Section 9.3, we will consider how to design a trading method based on the IMFs obtained from the EMD.

9.3 Trading Method Based on Empirical Mode Decomposition

The EMD is used in various financial applications. In Huang *et al.* (2003), Huang *et al.* use the EMD as a filter to extract variability of different scales. In Hong (2011), Hong derives an EMD-based forecast method for oil futures data with high frequency. Drakakis (2008) applies the EMD technique on the Dow-Jones volume and makes some inferences on its frequency content. In Zhu (2006), Zhu combines EMD and linear segment approximation to detect suspicious transaction. Guhathakurta *et al.* (2008) decompose the financial time series by EMD to analyze two different financial time series and compare their IMF phase and amplitude probability distribution. In this context, we focus more on the technical analysis aspect and aim to derive an EMD-based indicator for market entries.

Let $(x_1, x_2, \ldots, x_n)$ be the n-length realization of the index closing prices, and x_j be the index closing price on day j for $j = 1, 2, \ldots, n$. We will measure the profitability of the EMD-based indicator during this n-length period. Again we need to classify each day as either a buy, sell or neutral day. For each day $j = 1, 2, \ldots, n$, we perform the EMD algorithm for the following signal up to day j:

$$\mathbf{x}_j = (x_{-n_0}, \ldots, x_{-1}, x_0, x_1, x_2, \ldots, x_j),$$

where $n_0 + 1$ is the number of trading days in the previous year before the n-length observation period. The EMD gives the following decomposition for $\mathbf{x}_j$ at day j:

$$\mathbf{x}_j = \sum_{k=1}^{m^{(j)}} \mathbf{c}_k^{(j)} + \mathbf{r}^{(j)},$$

where

$$\mathbf{c}_k^{(j)} = \left(c_{k,-n_0}^{(j)}, \ldots, c_{k,0}^{(j)}, c_{k,1}^{(j)}, \ldots, c_{k,j}^{(j)}\right)$$

and

$$\mathbf{r}^{(j)} = \left(r_{-n_0}^{(j)}, \ldots, r_0^{(j)}, r_1^{(j)}, \ldots, r_j^{(j)}\right).$$

Note that the decomposition differs for each day j and the notations (j) are used to show that the decomposition, including the number of IMFs, depends on day j. We then design an indicator based on the above decomposition.

The idea is to construct two lines, the short line with higher frequency and the long line with lower frequency, to generate buying and selling signals. Recall that by changing the number of IMFs in the summation, we can derive components of different frequency range from the EMD. Therefore, we use two different partial sums of IMFs to construct the short line and the long line, such that we can replicate a momentum indicator. Firstly, at day j, we define the following value

$$EMDMA_{j_1,k_1,j} = \sum_{k=k_1}^{m^{(j)}} c_{k,j-j_1}^{(j)} + r_{j-j_1}^{(j)}, \quad j = 1, 2, \ldots, n.$$

As j varies from 1 to n over the n-length observation period, all the values $EMDMA_{j_1,k_1,j}$ will then form the short line EMDMA_{j_1,k_1}. Here, EMDMA is the short term for the combination of EMD and moving average. The parameter k_1 is selected as a small number (summing up most of the IMFs) to obtain a component of shorter cycle. The other parameter j_1 acts like a lagging parameter that makes use of an older value at time $j - j_1$ on the short line to stabilize our indicator. This idea stems from the numerical instability of the EMD algorithm. In particular, the end effects of the EMD algorithm spoil the applicability of the end values, because they change dramatically in response to even a single incoming closing price. Not only the lag index resolves the end effect problem, but also serves the purpose of creating crossovers with the long line defined next. Note that when $k_1 = 1$

and $j_1 = 0$, the short line is just reduced to the original data. Similarly, at day j, we then use the following value

$$EMDMA_{j_2,k_2,j} = \sum_{k=k_2}^{m^{(j)}} c_{k,j-j_2}^{(j)} + r_{j-j_2}^{(j)}, \quad j = 1, 2, \ldots, n,$$

where we use fewer IMFs in the summation with a $k_2 > k_1$, and correspondingly we also need a larger j_2 such that $j_2 > j_1$ to stabilize the long line, and create crossovers as a result. Over the n-length observation period, all the values $EMDMA_{j_2,k_2,j}$, $j = 1, 2, \ldots, n$ form the long line EMDMA_{j_2,k_2}.

Then, we use the short and long lines to design a trading strategy which is similar to the moving average crossover rule. A day j is assumed to be a buy day if

$$EMDMA_{j_1,k_1,j} > EMDMA_{j_2,k_2,j} \times \left(1 + \frac{p}{100}\right)$$

and a day j is a sell day if

$$EMDMA_{j_1,k_1,j} < EMDMA_{j_2,k_2,j} \times \left(1 - \frac{p}{100}\right).$$

Here a $p\%$ band is put around the long line to avoid frequent trades. We measure the profitability of this EMD-based indicator in Section 9.4.

9.4 Evaluation Results

We export the daily Hang Seng Index and China Shanghai Composite Index from Yahoo! Finance during the period from January 2007 to December 2011 as our test subjects. We measure the performance of the EMD-based indicator by studying the test statistics of profits generated by the buy and sell signals. We use the symbol $(k_1, j_1, k_2, j_2, p\%)$ to denote the EMD-based trading strategy, where the parameters k_1 and j_1 control the short line, k_2 and j_2 the long line, and p the band; see Section 9.3 for more details. We use the EMD algorithm designed in Rilling *et al.* (2003), where its MATLAB code is available online via the website [Rilling *et al.* (2007)].

Figure 9.2 displays the seven IMFs together with a residual for the Hang Seng Index from 2007 to 2011. The first few IMFs correspond to the high-frequency behavior of the index. The next few IMFs display the mid-range frequency components. For instance, the fifth IMF (its closeup is given in Figure 9.1) roughly represents a semi-annual cycle. The last few IMFs plus the residual represent the trend of the index. By changing the number of IMFs being summed, we derive a smoothened version of the

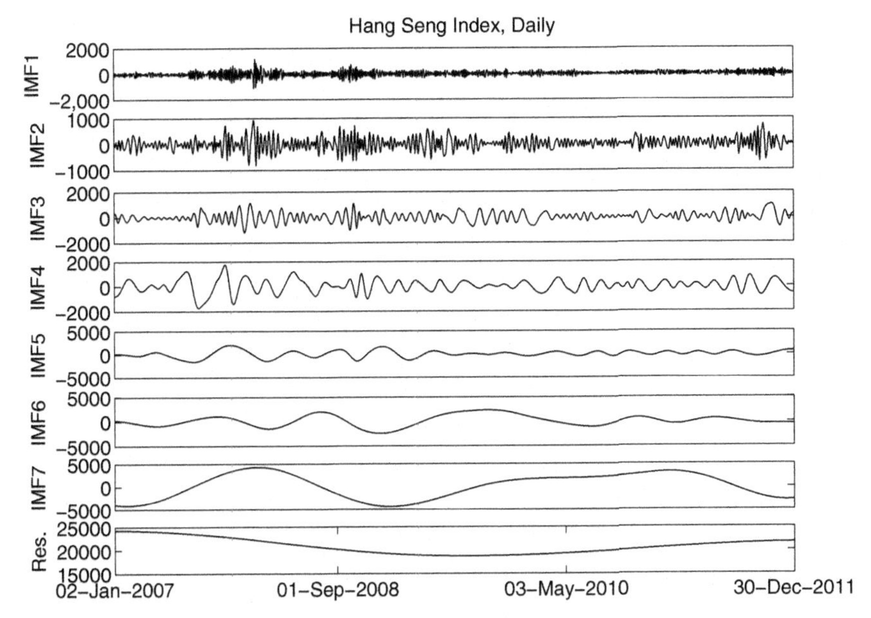

Figure 9.2: IMFs for the Hang Seng Index from 2007 to 2011 obtained through the EMD algorithm

original Hang Seng Index with different frequency ranges (see Figure 9.3 for an illustration).

Tables 9.1 and 9.2 report the mean daily return $\hat{\mu}_{\rm b}$ from the long period with the corresponding test statistic $T_{\rm b}$, the mean daily return $\hat{\mu}_{\rm s}$ from the short period with the corresponding test statistic $T_{\rm s}$, the mean return difference $\hat{\mu}_{\rm b} - \hat{\mu}_{\rm s}$ with the corresponding test statistic $T_{\rm bs}$, and the break-even costs C; see Section 7.2 for reference. In addition, the numbers of buy days and sell days, $N_{\rm b}$ and $N_{\rm s}$, generated by the EMD-based trading strategy are reported in the tables.

In Table 9.1, we find that the mean daily returns $\hat{\mu}_{\rm b}$'s for the long strategy are all positive, ranging from 0.012% to 0.111%. Overall, the EMD-based indicators generate profits similar to the moving average family in Hong Kong's stock market; see Table 7.1 for the corresponding results. Among the 10 EMD-based indicators, the rule (2, 5, 5, 100, 1%) gives the best mean daily return of 0.111%, or 27.97% annually. Table 9.1 also shows the performance of the short strategy generated by the EMD-based trading rules for Hang Seng Index. Their mean returns $\hat{\mu}_{\rm s}$'s are all negative.

In Table 9.2, the same set of EMD-based trading rules is applied to the China Shanghai Composite Index, during the period from 2007 to 2011.

Table 9.1: Standard test results and break-even costs for Hang Seng Index from 2007 to 2011, with EMD-based trading rules

Rule	N_b	N_s	$\hat{\mu}_b$ (T_b)	$\hat{\mu}_s$ (T_s)	$\hat{\mu}_b - \hat{\mu}_s$ (T_{bs})	C
(1, 0, 4, 50, 0%)	659	574	0.00012	−0.00030	0.00042	0.0038
			(0.23)	(−0.19)	(0.34)	
(1, 0, 4, 50, 1%)	605	529	0.00037	−0.00028	0.00061	0.0069
			(0.51)	(−0.13)	(0.47)	
(2, 5, 4, 50, 0%)	670	563	0.00058	−0.00086	0.00144	0.0198
			(0.77)	(−0.65)	(1.18)	
(2, 5, 4, 50, 1%)	608	522	0.00037	−0.00100	0.00137	0.0197
			(0.50)	(−0.73)	(1.05)	
(1, 0, 5, 100, 0%)	627	606	0.00058	−0.00076	0.00135	0.0218
			(0.76)	(−0.59)	(1.14)	
(1, 0, 5, 100, 1%)	606	571	0.00060	−0.00071	0.00131	0.0265
			(0.77)	(−0.53)	(1.07)	
(2, 5, 5, 100, 0%)	622	611	0.00102	−0.00120	0.00222	0.0760
			(1.22)	(−1.00)	(1.88)	
(2, 5, 5, 100, 1%)	607	574	0.00111	−0.00108	0.00219	0.0646
			(1.30)	(−0.86)	(1.78)	
(3, 15, 5, 100, 0%)	614	619	0.00080	−0.00095	0.00176	0.1084
			(0.97)	(−0.78)	(1.50)	
(3, 15, 5, 100, 1%)	601	587	0.00082	−0.00091	0.00174	0.0607
			(0.99)	(−0.72)	(1.43)	

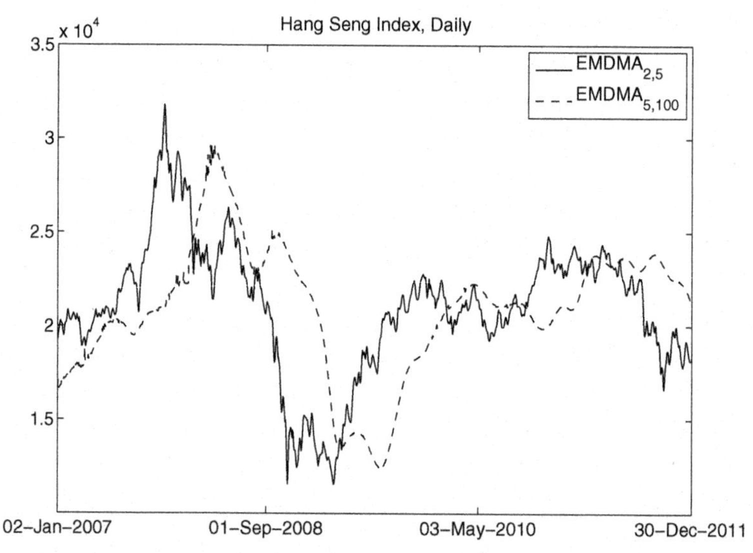

Figure 9.3: The short line and the long line generated by the EMD-based trading rule for the Hang Seng Index from 2007 to 2011

Table 9.2: Standard test results and break-even costs for China Shanghai Composite Index from 2007 to 2011, with EMD-based trading rules

Rule	N_b	N_s	$\hat{\mu}_b$ (T_b)	$\hat{\mu}_s$ (T_s)	$\hat{\mu}_b - \hat{\mu}_s$ (T_{bs})	C
(1, 0, 4, 50, 0%)	577	638	0.00147 (1.66)	−0.00166 (−1.46)	0.00312 (2.71)	0.0732
(1, 0, 4, 50, 1%)	556	619	0.00148 (1.66)	−0.00180 (−1.58)	0.00328 (2.79)	0.0745
(2, 5, 4, 50, 0%)	589	626	0.00141 (1.61)	−0.00166 (−1.46)	0.00307 (2.66)	0.1039
(2, 5, 4, 50, 1%)	565	606	0.00160 (1.77)	−0.00165 (−1.42)	0.00325 (2.74)	0.0792
(1, 0, 5, 100, 0%)	638	577	0.00126 (1.51)	−0.00176 (−1.50)	0.00303 (2.59)	0.0608
(1, 0, 5, 100, 1%)	610	553	0.00122 (1.44)	−0.00181 (−1.50)	0.00303 (2.51)	0.0498
(2, 5, 5, 100, 0%)	644	571	0.00117 (1.40)	−0.00168 (−1.42)	0.00285 (2.44)	0.0713
(2, 5, 5, 100, 1%)	621	549	0.00124 (1.45)	−0.00187 (−1.55)	0.00310 (2.57)	0.0854
(3, 15, 5, 100, 0%)	654	561	0.00119 (1.43)	−0.00177 (−1.49)	0.00296 (2.53)	0.0804
(3, 15, 5, 100, 1%)	624	540	0.00144 (1.67)	−0.00176 (−1.46)	0.00320 (2.66)	0.0974

Most of the results are better than those in Table 9.1 especially with higher significant levels T_b and T_s, echoing the inefficiency of the Chinese stock market observed in Tables 7.2 and 8.2. Among the 10 EMD-based indicators, the rule (2, 5, 4, 50, 1%) gives the best mean daily return of 0.160%, or 40.32% annually. For the short strategy generated by the EMD-based trading rules in the Chinese stock market, the mean returns $\hat{\mu}_s$'s are also all negative.

9.5 Summary

In this chapter, a new trading strategy based on EMD, the empirical mode decomposition, is presented. The strategy makes use of two different kinds of IMF summation to detect market entries and exits, much similar to the moving average crossover rule. Numerical results show that the tested EMD-based trading strategies outperform the moving average crossover rules in the Hong Kong and Chinese stock markets.

PART 3

Other Indicators in Technical Analysis

Chapter 10

Gann's Theory Unraveled[1]

10.1 Introduction to William D. Gann

William D. Gann (1878–1955) is a legendary market guru for the first half of the century. His most famous prediction is in the 1909 September wheat futures contract. Gann expects the contract to trade at $1.20 a bushel. On the last day of trading, with hours left to the expiration, Gann reasserts his belief. True to this expectation, the contact rallies from $1.08 per bushel to close exactly at $1.20 per bushel, confirming his forecast. Yet much of his work is shrouded in mysticism. In 1976, Billy Jones purchased Lambert–Gann Publishing Company from Gann's former partner, Edward Lambert, and reprinted much of Gann's original work which sparkles a new era of technical analysis for its faithful practitioners.

10.2 Gann's Theory Unraveled

Gann's theory [Gann (1935); Reddy (2012)] suggests that there is a proportional relationship between price and time. Simply explained, time must be allowed to lapse before prices can reverse direction. The concept of squaring of price and time is introduced here. This occurs when a unit of price equals a unit of time. This is a perfect state of balance between time and price as is depicted by the 45-degree line. This is the famous "Gann line" drawn from a historical top or bottom to determine the market movement. The upward Gann line drawn from a market low acts as a support line in a

[1]This chapter is mainly based on the work [Chew *et al.* (1996)] by one of the authors with updated examples.

bull market, while the downward one drawn from a market high acts as a resistance line in a bear market.

10.3 Gann's Geometric Angles

The Gann line represents Gann's major upward or downward trend line. A breaking of this line represents a major trend reversal. Aside from the 45-degree Gann line, there are eight other important angles altogether in Gann's analysis. Lines with these significant angles are all known as Gann's trend lines.

time × price	degree
1×8	82.5
1×4	75
1×3	71.25
1×2	63.75
1×1	45
2×1	26.25
3×1	18.75
4×1	15
8×1	7.5

As mentioned, the 45-degree Gann line represents a perfect balance of price and time. This implies that a unit of price is equaled to a unit of time. In the following, it is represented by the notation 1×1, which means prices increase or decrease at the speed of time. The next steeper line is the 1×2 line which implies prices increase or decrease of two price units for each unit of time. Gann's trend lines with different significant angles may act as trends (as the name already implies), channels, support or resistance in trading. Upward Gann's trend lines are graphically represented in Figure 10.1. These lines will be drawn from a market low in practice. Similarly we can draw downward Gann's trend lines starting from a market high.

To draw Gann's trend lines, one first chooses a price unit and a time unit. For example, the time unit is typically one day. However, the price unit is usually not 1 point, and needs to be scaled to go along with the time unit. After that, Gann's trend lines with certain specific angles are drawn from market tops or bottoms. For example, in the downtrend, the most important line is the one with angle $1 \times (-2) = -63.75$ degree, which represents the situation when prices decrease of two price units for each unit

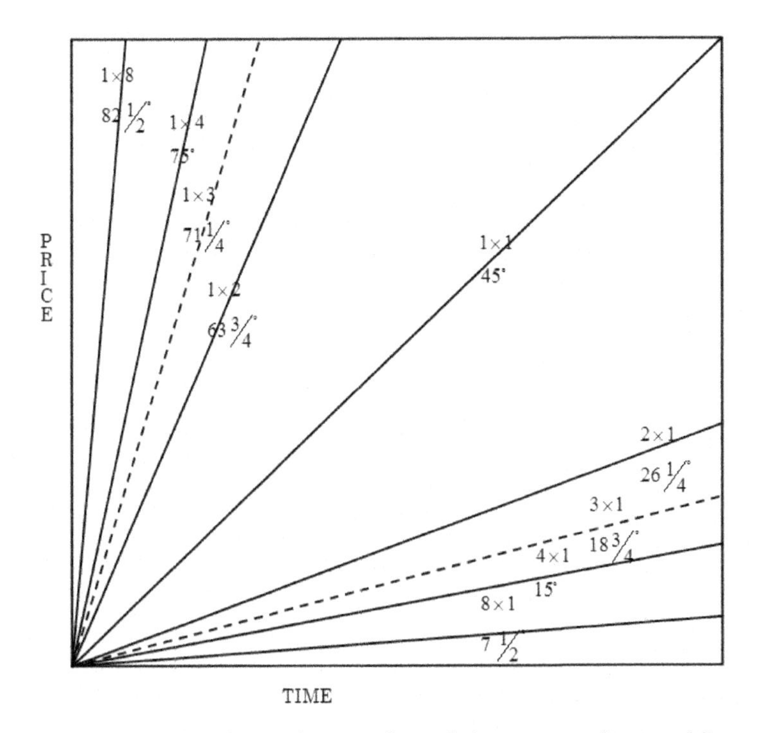

Figure 10.1: Gann's significant angles and the corresponding trend lines

of time. For convenience, we skip all the minus signs (−) for the degree in the downtrend in this context.

10.4 Gann's Retracement Percentages

Gann (1935) also believes that there is a natural tendency for historic highs and lows to determine support or resistance areas. He especially stresses the importance of the 50% retracement, which is the mid-point of a certain market high and market low. This is invariably linked with the twelfth century Italian mathematician, Leonardo Fibonacci, who also places similar emphasis on the rule that the universe is governed by a precise mathematical pattern.

We shall now discuss Gann's horizontal retracement method, which is used along with Gann's trend lines. The method is actually a similar tool to Fibonacci retracement [Boroden (2008)]. In classic technical analysis, Fibonacci retracements are formed by dividing the price action between two extreme points into Fibonacci ratios. Then, horizontal lines are drawn at these levels to indicate support or resistance. In Gann's theory, it is believed that the 50% retracement is crucial, which happens to be a Fibonacci ratio as well. Instead of Fibonacci ratios, Gann divides the price action into eighths: $\frac{1}{8}$, $\frac{2}{8}$, $\frac{3}{8}$, $\frac{4}{8}$, $\frac{5}{8}$, $\frac{6}{8}$, and $\frac{7}{8}$. The ratios, as derived respectively, are 12.5%, 25%, 37.5%, 50%, 62.5%, 75%, and 87.5%. Like the Soviet economist Nikolai Kondratiev [Kondratiev (1925)], Gann believes in full cycles in the sense that it is possible for prices to fully retract after a prolonged bull period. To Gann, the anniversary of, say, the October crash in 1987 is important.

After the 50% retracement, the next two, in order of significance, are the 37.5% and the 62.5% retracements. The 62.5% one is close to one of the Fibonacci ratios 61.8%, which is also called the key Fibonacci ratio in technical analysis for its importance. These two are followed by the 33.3% ($\frac{1}{3}$) and the 66.7% ($\frac{2}{3}$) retracements, which are the minimum and maximum benchmarks used by most market technicians. Gann's percentage retracement is often used in evaluating the Elliott wave principle [Elliott (1938)]. To extend Gann's idea, one may include the following retracements: $\frac{1}{6}$ (16.7%), $\frac{5}{6}$ (83.3%), $\frac{1}{5}$ (20%) and $\frac{4}{5}$ (80%).

We note that in Gann's theory the major turning points for stock prices are likely to happen in the intersection of two different trend lines or the intersection of a trend line and a retracement line.

10.5 Applications of Gann's Trend Lines and Retracements

Gann's trend lines and retracements have been seen as an effective yet interesting technical analysis tool that provides very good insight for the investor. We show that such techniques have been already applied successfully in the 90s via the following example of Straits Times Industrials Index (STII) in Singapore.

Application to STII

Note that the STII is the predecessor of the current Straits Times Index (STI), and STII includes the industrial categories back then. The STII is presented in Figure 10.2. In the figure, we construct the Gann's upward trend lines at a market low before the bull run in 1993 at point A, the Gann's downward trend lines at a market high at point E, and the corresponding $\frac{1}{8}$, $\frac{2}{8}$, $\frac{3}{8}$, $\frac{4}{8}$ and $\frac{5}{8}$ retracement lines.

From the figure, we find the following:

(1) The 1993 bull run is well supported by the 45-degree trend line which is also the strong trend line for the bull run in 1994.

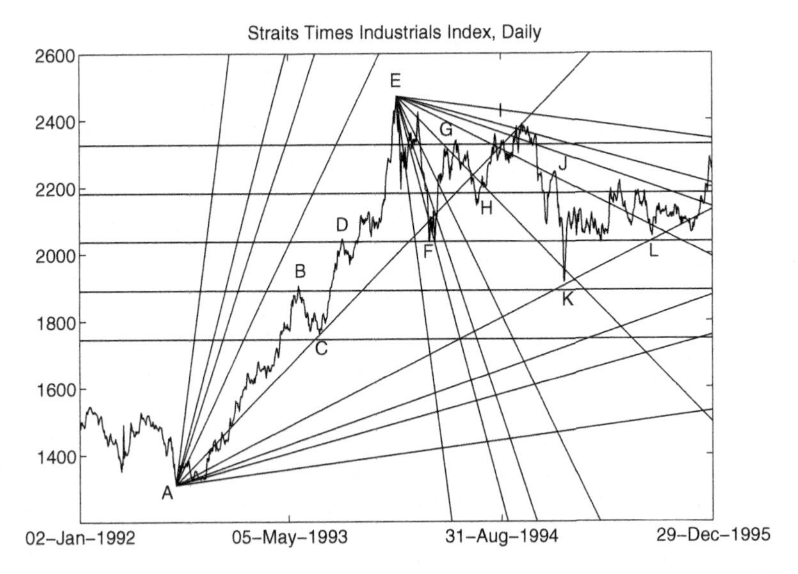

Figure 10.2: STII with its Gann's trend lines and retracements

(2) The 1993 bull run first stops at point B on 31 May 1993, retraces to point C on 28 June 1993 and runs to point D on 7 September 1993 before it goes to a market high at point E on 7 January 1994. Point B is on the midpoint of the whole bull run from points A to E. Thus, point B actually gives us a clue as to what the market high will be.

(3) The distance from point B to point C is the same as the distance from point B to point D, the next peak. The $\frac{1}{8}$ retracement distance will enable us to forecast the correction of STII after it reaches point D.

(4) The bear market starts from point E and the downtrend in the first half of 1994 is well supported by the $\frac{3}{8}$ retracement and the 45-degree upward trend line from point A.

(5) From point F to the end, the $\frac{1}{8}$ retracement line is a strong resistance line and the $\frac{3}{8}$ retracement line is a strong support line, while the $\frac{2}{8}$ retracement line is the support line from points G to J and becomes the resistance line thereafter.

(6) The 45-degree downward trend line from point E is a strong resistance line for STII from points E to H and becomes a strong support line thereafter.

(7) Point F touches the $\frac{3}{8}$ retracement line, near to the intersection of the 45-degree upward trend line and the 75-degree downward trend line. Similar situations can be found near points I, K and L. For instance, it is interesting to note that STII dips at 1919.2 (point K) on 24 January 1995. Point K touches the $\frac{4}{8}$ retracement line, near to the intersection

of the 26.25-degree upward trend line and the 45-degree downward trend line.

(8) From point K onwards, the 26.25-degree upward trend line from point A becomes a strong support line for STII. STII rebounds at points K and L once it touches the trend line.

(9) On 16 November 1995, STII crosses the 26.25-degree trend line, dives to the south and rebounds back again. It then breaks through all the way to above the $\frac{1}{8}$ retracement line as well as the 7.5-degree line from E. These give good signs to a bull run in STII. Though the previous high at point E is likely to be a resistance, the STII breaks through this point and continues the rise all the way to 2484 on 5 February 1996.

Nowadays, we can still discover relevance of Gann's theory in application to indices or stock prices. Next, we shall apply Gann's idea to analyze the Standard & Poor's 500 (S&P 500). To demonstrate the theory to individual stocks, we choose the HSBC Holdings listed in Hong Kong's stock market.

Application to S&P 500

The Standard & Poor's 500 (S&P 500) is presented in Figure 10.3. In the figure, we construct the Gann's upward trend lines at the lowest point before

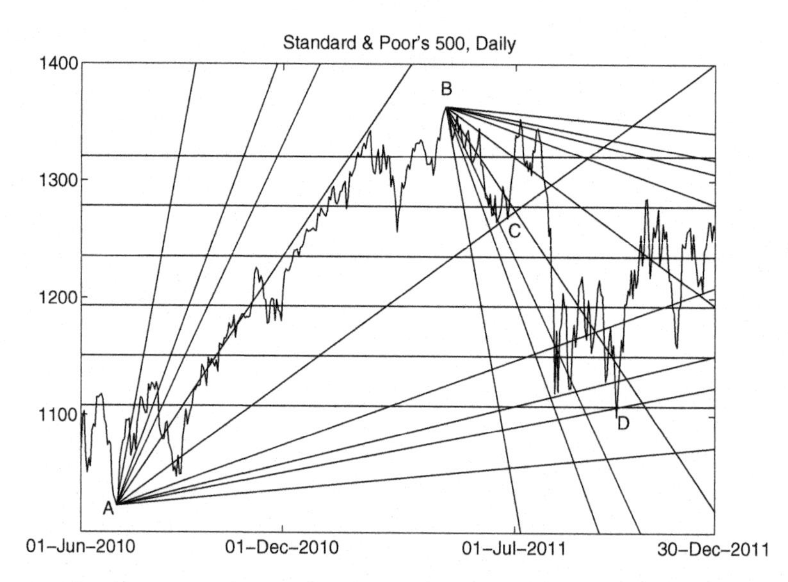

Figure 10.3: S&P 500 with its Gann's trend lines and retracements

the bull run in 2010 at A, the Gann's downward trend lines at the highest point B, and the corresponding $\frac{1}{8}$, $\frac{2}{8}$, $\frac{3}{8}$, $\frac{4}{8}$, $\frac{5}{8}$ and $\frac{6}{8}$ retracement lines.

From the figure, we find the following:

(1) The 2010 bull run goes along with the 63.75-degree upward trend line which also acts as a strong resistance line for the bull run in late 2010 and the first half of 2011.
(2) The bull run reaches the highest point at B on 29 April 2011. The 45-degree upward trend line is a strong support line before the market suffers the sudden drop in August 2011.
(3) The downtrend begins at point B and it is well supported by the $\frac{6}{8}$ retracement and the 15-degree upward trend line from A. It is also mildly supported by the 18.75-degree upward trend line.
(4) The $\frac{2}{8}$ retracement line is a strong support line before the drop in August 2011 but then it becomes a strong resistance line thereafter.
(5) The index finds strong support at where the 45-degree upward trend line intersects the 63.75-degree, 71.25-degree, and 75-degree downward trend lines near point C.
(6) The index rebounds significantly after point D, the intersection of the $\frac{6}{8}$ retracement line and the 15-degree upward trend line.

Application to HSBC Holdings

Gann's theory can also be used to analyze individual companies. We choose HSBC Holdings during the period 2009–2011 and present it in Figure 10.4. We construct the Gann's upward trend lines at the lowest point before the bull run in 2009 at A, the Gann's downward trend lines at the highest point C, and the corresponding $\frac{1}{8}$, $\frac{2}{8}$, $\frac{3}{8}$, $\frac{4}{8}$ and $\frac{5}{8}$ retracement lines.

From the diagram, we find the following:

(1) The first leg of the 2009 bull run goes along with the 82.5-degree upward trend line from point A.
(2) The 75-degree upward trend line is a strong support line for HSBC Holdings for the 2009 bull run.
(3) The distance between the $\frac{1}{8}$ and the $\frac{2}{8}$ retracement lines implies the vertical rise from the $\frac{1}{8}$ retracement to the highest point B. Therefore it actually gives us a clue as to what the highest point would be.
(4) After HSBC Holdings peaks at point B, it dives and the $\frac{1}{8}$ retracement line becomes its strong resistance line.
(5) After point B, the 75-degree downward trend line from point B becomes a strong support line.

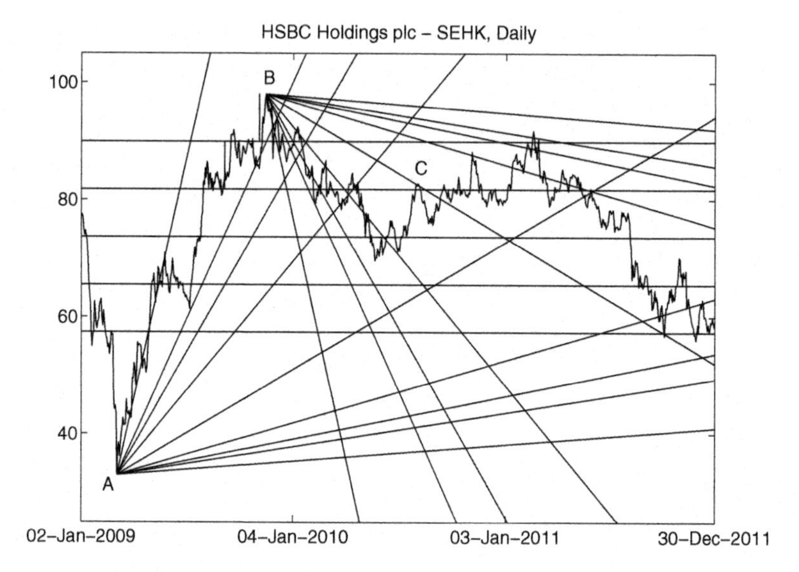

Figure 10.4: HSBC Holdings with its Gann's trend lines and retracements

(6) The 45-degree downward trend line intersects the $\frac{2}{8}$ retracement line at about the price level \$81.75 at point C. The HSBC Holdings keeps coming back at this price during 2010 and the first half of 2011.

(7) The $\frac{5}{8}$ retracement line seems to be a support line for HSBC Holdings near the end of 2011.

10.6 Summary

As can be seen from the results, Gann's theory does give us some very powerful and interesting trading signals. It also offers remarkable insight into the workings of the market as well as enormous flexibility in its use. It is amazing how the market seems to bounce off just at or near the retracement lines and trend lines. It is also relatively simple to use, especially if one has a computer software package to use along with it. What is more, it can be used on indices as well as individual stocks.

Chapter 11

Bollinger Bands

11.1 Introduction to Bollinger Bands

Bollinger bands belong to a type of trading instruments called the envelope. Envelopes are essentially bands around some indicators like the moving average. They are in general effective for filtering out the whipsaws that are inherent in trend following indicators. Envelopes can be made to be very simple or extremely complicated. A simple example is a single moving average surrounded by a band that is a certain percentage (or sometimes a certain fixed amount) away from the moving average. These bands will contain the normal price movements, especially during non-trending ranges. When the price breaks out of the bands, it usually indicates that a trend is about to start. When the price goes back into the bands again, it is usually indicative of the end of a trend.

There are many ways to vary the envelopes. Firstly, the indicator around which the envelope forms can be almost any indicator. It can be, for example, simple moving average (SMA), exponential moving average, or moving average convergence–divergence (MACD). In this context, we will only use the moving average for illustration. The distance of the lower bands from the indicator can also be determined in many ways, and the criteria for plotting upper and lower bands can be different. For example, in a trending market, the upper band can be two standard deviations above the indicator while the lower band is one standard deviation below it. Neither need the bands be based on the same price variable as the indicator. For example, the indicator can be based on the closing price, the upper band the daily high, and the lower band the daily low.

In light of all this, Bollinger bands are a relatively new type of envelope. They are initially known as alpha–beta bands and are popularized by John Bollinger, a market analyst for CNBC/Financial News Network. It is essentially a moving average with a *moving standard deviation* around it. Hence, the bands will be wider in volatile markets and narrower in quiet ones.

11.2 Calculation of Bollinger Bands

The first thing to do is to plot the moving average. In the case of Bollinger bands, the recommended period is 20 days, but this can be varied as the need arises. The standard deviation needs to be determined next. In general, the standard deviation is a measure of volatility and can be calculated at time j as follows:

$$s_j = \sqrt{\frac{1}{M}\sum_{k=0}^{M-1}(x_{j-k} - SMA_{M,j})^2},$$

where M is the time period, x_j is the price at time j, and

$$SMA_{M,j} = \frac{1}{M}(x_j + x_{j-1} + \cdots + x_{j-M+1})$$

is the M-period SMA at time j. For Bollinger bands, we use the recommended period of 20 days and let $M = 20$.

The usual norm is to place the bands two standard deviations away from the moving average. Bollinger's explanation is that two standard deviations would encompass most of the subsequent price movements. As can be seen from the formula, the deviation from the average price is squared. This makes it very responsive to the short term price changes, expanding and contracting in accordance with recent market movements. Although standard deviation is a statistical concept, there should not be any statistical assumptions beyond the observation that the price tends to stay within the bands. Of course the norm of using a 20-day period and two standard deviations needs not be strictly followed, but should be modified to suit needs.

11.3 Trading Methods Using Bollinger Bands

There are many ways of trading with envelopes, but we shall outline only the more commonly used methods.

Trading Methods in Trending Market

The traditional method of using envelopes is to use it as a trend following device, much like the moving average or the MACD. Here, the main indication of a trend starting is the penetration of a band. The method then is to enter the market in the direction of the breakout, i.e., buy when the upper band is penetrated, or sell when the lower band is penetrated (if short selling is allowed). Having entered the market, there are two methods of timing the exit:

(1) exit when the opposite band is penetrated;
(2) exit when the moving average between the bands is reached.

Either set of rules will ensure that the major trends will not be missed. In a market where short selling is allowed, the first method will be a reversal system. Investors will be long one moment and short the next moment when the other band is penetrated. The second method is slightly different in the sense that if the price stays within the band after the position has been liquidated, there will be no position held. This is a neutral zone, and trade will only begin if there is another breakout. The other good point about the second method is that the drawdown is limited to the difference between the band and the moving average instead of the difference between the two bands, as would have been the case for the first method.

Figure 11.1 features the chart for HSBC Holdings with its Bollinger bands. It is produced by using the MATLAB command `bollinger`. It demonstrates the first method, i.e., exiting trades only when the opposite band is penetrated. The first entry, as indicated at point A, is in April 2009 when the upper band is penetrated.

The price first moves around the upper part of the bands along with the bull run and then starts its consolidation. By July 2009 (point B), the lower band is penetrated and the position is liquidated. Shortly after, the upper band is again penetrated at point C. After this line, the price goes through a steep climb before bending down a few times below the moving average between September 2009 and December 2009. During this period, the price never actually touches the lower band. However, if the second method of exiting trades when the price touches the moving average has been used during this period, a more profitable scenario will have resulted. The lower band is then penetrated at point D, in late January 2010. Up till this point, the system has managed to capture the two main trends and in doing so succeeds in making a substantial profit. The next two trades are executed at a time when there is no clear trend. The third long trade starts

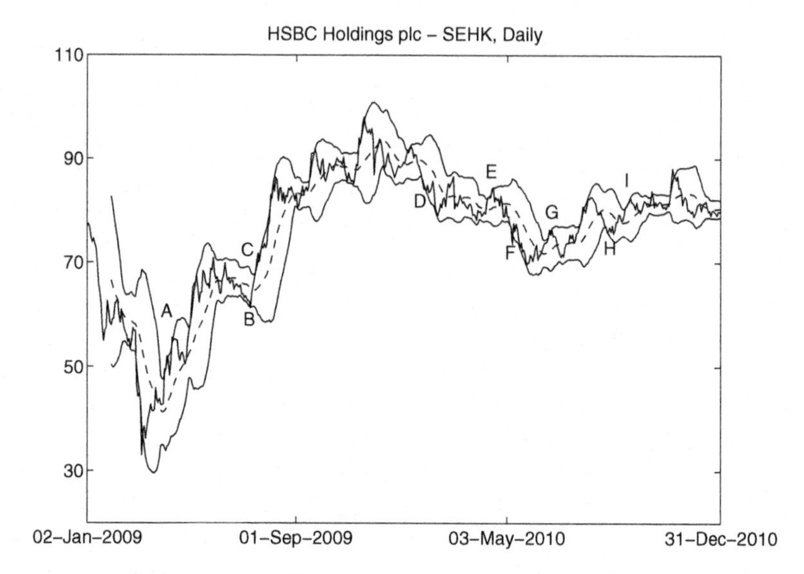

Figure 11.1: HSBC Holdings with its Bollinger bands

at point E and ends at point F, resulting in a loss. Since this is a reversal system, a short position would have been taken if short selling is allowed. This is immediately followed by an entry at point G and an exit at point H, which results in an ignorable profit. The last entry is at point I, and seems to be not very promising.

We can see that the first trading method of using Bollinger bands is very profitable in a trending market. In a non-trending market, this approach frequently results in whipsaws, as can be seen from the trades between points E and I in Figure 11.1. We shall come back to this matter after illustrating the second trading method involving Bollinger bands in a trending market.

Figure 11.2 features the Nikkei 225 with its Bollinger bands and MACD. This time we add the MACD for comparison. In this case, we are using the second exit method, that is, exiting when the closing price touches the moving average. The first entry is at point A. In comparison, the MACD also gives a buy signal at about the same time. Note that the MACD gives a buy signal when the MACD line (the solid line) penetrates the signal line (the dashed line) from below, and a sell signal when the penetration is from above. The sell signal is at point B where the closing price touches the moving average. The MACD also gives a sell signal near this point. A buy signal is given again at point C, with the accompanying sell signal at point D. This set of signals also happens to coincide with the signals given

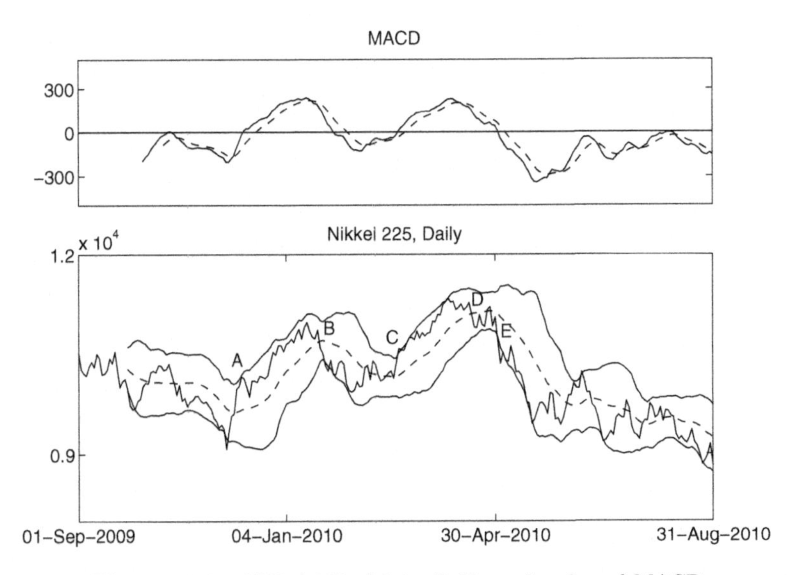

Figure 11.2: Nikkei 225 with its Bollinger bands and MACD

by the MACD. The last signal is a sell signal at point E. This is in fact a signal to sell short, and judging from the downtrend after point E, this signal is fairly accurate. It is interesting to note that the point E coincides with the point where the MACD line crosses the zero mark.

In general, the traditional method of using Bollinger bands as a trend following device enables the capture of major trends. It is in fact comparable to the MACD in generating useful trend following signals. Care must still be taken when using the bands as they sometimes do give false signals. For instance, in Figure 11.1, we will get a false signal right after point A if we apply the second method of exiting. Better trading decisions can be made if Bollinger bands are used together with other indicators.

Trading Method in a Non-trending Market

Since it is generally observed that the price usually stays within the bands in a non-trending market, it is only natural that a trading method can be devised to trade within the band. This method effectively uses Bollinger bands as a counter-trend trading method, or an overbought/oversold indicator much like the relative strength index (RSI). Of course, as with other counter-trend methods, it is effective only in non-trending markets. In trending markets, as we have already seen, the price tends to go beyond the bands and stay there.

The trading method is relatively simple, just buy when the price touches the lower band. If the price moves against you and penetrates the lower band, it may be an indication that a downward trend has started. Therefore, it is advisable to exit quickly and take a small loss. If the price movement after entry is favorable, which is more often the case, hold on until the price touches the upper band. The position is then liquidated, and a short position is entered, if short selling is allowed. The same rules for cutting losses and reversing trades are subsequently followed.

In Figure 11.3, we have a chart of Industrial and Commercial Bank of China (ICBC) with its Bollinger bands. The particular time period chosen coincides with a period when the market is non-trending. Hence, application of the counter-trend technique using Bollinger bands is particularly well suited. The first point of entry is at point A, where the price touches the lower band, giving a buy signal at about \$5.40. Following this, the price reaches the upper band again at point B, giving a sell signal at about \$5.90. If short selling is allowed, a short position should have been taken at this point, and subsequently covers up at point C, at about \$5.65. At point C, more shares would have been bought, and be liquidated at point D around \$6.30. As can be seen from Figure 11.3, such signals can be extremely profitable. In fact, the gap between these two sets of buy/sell prices is about \$0.50. During this period, there is one more trading opportunity at points E and F, with the gap between buying and selling in the range of \$0.50.

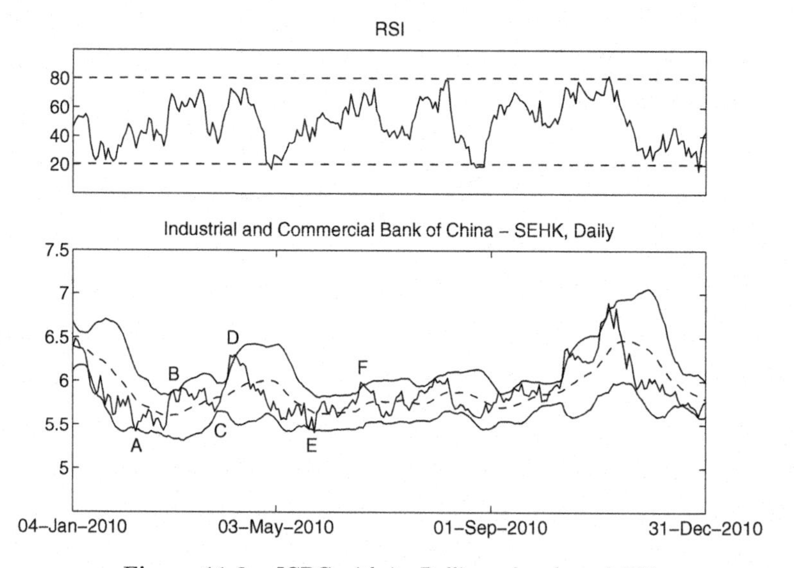

Figure 11.3: ICBC with its Bollinger bands and RSI

This trading method of employing the Bollinger bands is comparable with another counter-trend method, the RSI. The RSI is an indicator used in technical analysis for measuring the strength of the market dynamics. It is proposed by J. Welles Wilder [Wilder (1978)]. The formula of RSI will be given in detail in Section 12.4 when we discuss the technical indicators based on RSI. Simply put, the RSI is a value between 0 and 100. Moreover, when the RSI hits the marks of 30 and 70, the market is said to be oversold and overbrought, respectively.

In Figure 11.3, we compare the RSI to the Bollinger bands for the ICBC example. The RSI is produced using the MATLAB command `rsindex`. Judging from the RSI range in Figure 11.3, we shall assume that the market is oversold when the RSI touches 20 and that the market is overbought when the RSI touches 80. As can be seen from Figure 11.3, near points A and C, where the Bollinger bands give a buying signal, the RSI is indeed at its relatively low points, respectively. However, the RSI has not given an exact oversold signal because it is still slightly above 20. The RSI does give the oversold signal near the end of April 2010, but the buying signal given by the Bollinger bands at point E has a lower stock price. Similarly, the Bollinger bands give the selling signals at points B, D, and F, but the RSI fails to give overbought signals. The RSI does not peak near these points, not to mention hitting the mark of 80. In this case, we see that the Bollinger bands are more sensitive than the RSI.

Through this example, we observe that a trading method using the Bollinger bands as a counter-trend tool can be very profitable in a non-trending market. In comparison to the RSI, it can also be more sensitive. This is partly due to the fact that the Bollinger bands take the volatility of the market into account, whereas the RSI does not.

In conclusion, it is of paramount importance to note if the market is trending before applying Bollinger bands in trading. One remarkable method of seeing whether the market is trending is to use the average directional movement index (ADX). The ADX is a technical indicator also developed by J. Welles Wilder in his famous book [Wilder (1978)]. The definition of ADX requires a set of some other indicators and will be discussed in detail in Section 12.5. Roughly speaking, ADX makes use of each day's high, low, and closing prices and has a value between 0 and 100. The ADX values of 20 and 40 are the benchmarks of trend weakness and strength respectively. However, one can also specify the values to meet various purposes. For instance, we could just take it that if the ADX is above 25 and rising, the market is deemed to be trending. Otherwise, the market is non-trending.

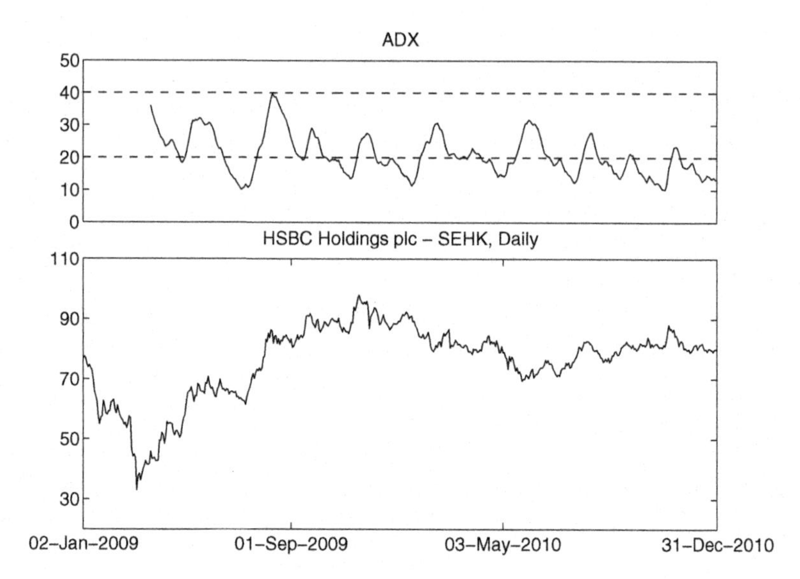

Figure 11.4: HSBC Holdings with its ADX

In Figure 11.4, we include the ADX of HSBC Holdings in the chart. We note that in the first half of the period, the ADX tends to cross the benchmark value 20 and then rises further away from it for some time. Hence, the market is mostly trending, and the trading method with Bollinger bands in a trending market should be applied; see Figure 11.1. In the second half of the period, the ADX lacks the power to rise way above 20 for most of the time. Therefore, the market is non-trending. A cautious trader will employ some non-trending technical indicators and the profits can still be quite substantial.

Volatility Squeeze

One advantage of using Bollinger bands is that it encompasses variable volatility in its analysis. Hence, the band width is wider in a volatile market, and it is narrower in a quiet market. A trading method of using this special feature is to note any sudden increase in volatility. When the market is subdued, the bands will narrow due to decreased volatility. Then interest picks up, the volatility will increase, and the bands will widen. This is an excellent time to enter the market in the direction of trend.

In Figure 11.5, Want Want China Holdings Ltd. is featured with its Bollinger bands. We note that there is a volatility squeeze at point A. We apply the technique of trading in the direction of the trend after the squeeze.

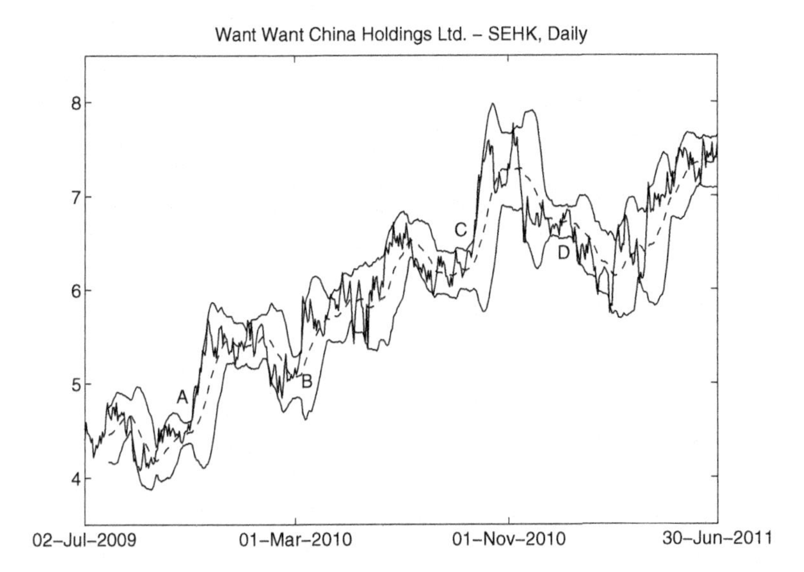

Figure 11.5: Want Want China Holdings Ltd. with its Bollinger bands

In this case, a long position is taken, after which the price goes up all the way to about \$5.70. The second squeeze is at point B, after which only a short rally takes place. The third squeeze is at point C, after which the price again shoots up to \$7.60 from around \$6.40. At point D, a squeeze occurs again, only this time the market is in a downtrend. A short position could have been taken if short selling is allowed. It must be remembered that this method gives only the entry signal. An alternative exit signal must be sought. If the exit signal is inappropriate, this method could turn out to be a disaster. If, on the other hand, a good exit signal is used, this method could be very profitable.

11.4 Summary

In this chapter, we have seen the flexibility with which Bollinger bands can be used. In the traditional sense, it can be used profitably as a trend following device. In fact, using the Bollinger bands in this way can have most of the major trends captured. It is comparable with the MACD in this respect, and at times even better. When used in a non-trending market, it is still shown to be immensely profitable. Compared to the RSI, it is a more sensitive indicator, generating more correct signals.

Care must be taken when using the Bollinger bands, and the nature of the market must be understood. Trying to use the Bollinger bands either as a trend follower in a non-trending market or as counter-trend instruments in a trending market would be disastrous. As such, one should incorporate the use of a trend detecting indicator like ADX when making the decision about whether to use Bollinger bands in a trend following manner or in a counter-trend manner. We also find that trading after periods of volatility squeeze can also be significantly profitable. However, this depends a lot on the chosen exiting method.

Finally, we note that one would be able to make better trading decisions if other indicators could be used in conjunction with Bollinger bands.

Chapter 12

Other Technical Indicators[1]

12.1 Standardized Yield Differential Indicator

Various studies have shown that fundamental data like business conditions, dividend yields, economic variables, and the popular price-to-earnings (P/E) ratios can predict stock returns to a large degree [Campbell (1987); Breen *et al.* (1989); Cochrane (1991)]. As a result, it is common to use economic or fundamental variables to create technical indicators and trading rules. In Wong (1993, 1994), Wong has introduced such kind of indicator, named by the standardized yield differential (SYD), which utilizes the difference between the earnings-to-price (E/P) ratio and the bond yield or the interbank interest rate. Ariff and Wong (1996) perform a linear regression on the SYD and asset prices, and conclude that there is a significant relationship between them.

The SYD is a monthly indicator computed from E/P ratios and bond yields. In this section, we use j to denote a time variable for the jth month specifically. Notice that the E/P ratio is the reciprocal of the P/E ratio. The E/P ratio at time j, denoted by EP_j, is a measure of market reaction to the earnings of all the firms in each stock market, calculated using the formula:

$$EP_j = \frac{E_j}{P_j} \equiv \frac{\sum_{k=1}^{m} w_{k,j} E_{k,j}}{\sum_{k=1}^{m} w_{k,j} P_{k,j}}, \tag{12.1}$$

[1]Section 12.1 is mainly based on Wong *et al.* (2002) by one of the authors.

where $E_{k,j}$ is the average earning per share for stock k at time j, $P_{k,j}$ is the average stock price for stock k at time j, $w_{k,j}$ is the weight of the stock k in the corresponding index, and m is the number of stocks in the index.

We remark that the E/P ratio $(= E_j/P_j)$ at time j is different from the more common earning yield $(= E_{j+1}/P_j)$ at time j. The former does not include the market expectation of earnings growth while the latter does; see [Brealey and Myers (1991)] for reference. However, the E/P ratio is based on openly available information and it is a common measure for the real equity earning relative to the equity price. The earning yield E_{j+1}/P_j data are not accessible to technical analysts at time j, so they are not employed in technical analysis. Hence, the E/P ratio is incorporated in the calculations for defining a technical indicator.

The monthly yield differential at time j, denoted by YD_j, is defined as

$$YD_j = EP_j - BY_j,$$

where EP_j is defined in (12.1) and BY_j is the bond yield or interest rate at time j. The SYD at time j over M months, denoted by $SYD_{M,j}$, is then calculated as

$$SYD_{M,j} = \frac{YD_j - \overline{YD}_{M,j}}{\sqrt{\frac{1}{M-1}\sum_{k=0}^{M-1}(YD_{j-k} - \overline{YD}_{M,j})^2}}, \tag{12.2}$$

where $\overline{YD}_{M,j}$ is the M-period simple moving average (SMA), or the mean of the monthly yield differential

$$\overline{YD}_{M,j} = \frac{1}{M}\sum_{k=0}^{M-1} YD_{j-k}.$$

The formula (12.2) shows that $SYD_{M,j}$ is a standardized measure of the SMA of the monthly yield differential. The smoothing period M in SYD is usually chosen from 24 to 36 months to capture a relatively long period [Wong *et al.* (2002)]. However, one may want to take a longer smoothing period like $M = 60$ months to catch the effect of the long run if he or she thinks that the bull market has been lasting for too long a period. According to Wong *et al.* (2002), a large SYD value indicates that

(1) yield differential is large with respect to the mean monthly differential $\overline{YD}_{M,j}$ and
(2) the yield from equity is comparatively higher than bond yields.

SYD is not used to find an accurate trend in the stock market or to forecast the economy. The application and interpretation of the SYD indicator in the stock market depends heavily on the investors under different market conditions. Two possible scenarios on using the SYD indicator are listed in the following:

- Scenario A: If the yield differential YD_j at time j is large compared with the mean monthly differential $\overline{YD}_{M,j}$, then SYD will become a large value. The reason could either be stock market correction, corporate profit increase, or bond/cash yield decrease. These factors are bound to happen during bullish periods for equities. Therefore, a large positive SYD value implies that stock prices are expected to rise and an investor has to pay for the stock investment. On the contrary, a large negative SYD value implies that the stock prices are likely to fall.
- Scenario B: Until higher earnings are reported, a decreasing E/P ratio reflects expectations of stronger economic prospects, which drive up bull runs. Therefore, a high E/P ratio may signal weaker economic prospects or the lack of confidence in an enterprise's future earnings. Compared to Scenario A, Scenario B has a totally opposite situation as a large positive SYD value implies that stock prices are likely to fall, and a large negative SYD implies that stock prices are likely to rise.

To fully utilize the SYD indicator, one needs to figure out whether Scenario A, Scenario B, or other possible scenarios is dominant in the market. Since the market is a combination of various scenarios, the SYD strategy can be subjective, and different market analysts can apply the SYD in their preferred ways.

To illustrate the trading method based on SYD, we take Scenario A as an example because Scenario B requires a wider range of economic variables. In Scenario A, a positive and large SYD value implies, there is rising power in the future price movement, while a negative and large SYD value implies a decline in future price movement. Using these implications, one may consider different values of SYD as market entry/exit points. For instance, one can regard SYD values greater than $+2$ (less than -2) as strong buy (sell) signals and SYD values between 0 and 2 (between -2 and 0) as weak buy (sell) signals.

12.2 52-week High Momentum Strategy

The 52-week high momentum strategy exploits the 52-week (1-year) high, a specific piece of price information, for making market entry decisions

[George and Hwang (2004)]. The idea stems from the observation that traders consistently underreact or overreact to good news hitting the market. Therefore, when prices start to climb close to the 52-week high from some recently arrived good news, trading opportunities appear because of the momentum created from the biases in traders' reactions.

For the 52-week high strategy, stocks are first ranked based on the nearness to the 52-week high price, which is defined by a ratio

$$\frac{x_{k,j}}{x_{k,j}^h},$$

where $x_{k,j}$ is the price of stock k at the end of month j and $x_{k,j}^h$ is the highest price of stock k during the 52-week period that ends on the last day of month j. Long or short positions are then taken in those stocks closer to or further away from the 52-week high.

In George and Hwang (2004), the 52-week high momentum strategy is examined using the CRSP US stock database from the period 1963 to 2001, and it is also compared with two other momentum investment strategies in Jegadeesh and Titman (1993); Moskowitz and Grinblatt (1999). The momentum strategy in Jegadeesh and Titman (1993) first ranks the individual stocks by their past return performances, and then traders long the top 30% stocks or short the bottom 30% stocks. The momentum strategy in Moskowitz and Grinblatt (1999) first ranks the industries by their past return performances, and investors will take long position in the top 30% industries or short position in the bottom 30% industries. The returns from the 52-week high momentum strategy double the size of those from the other two momentum strategies [George and Hwang (2004)]. The performance of the 52-week high momentum strategy is justified in George and Hwang (2004). The main reason is that the 52-week high price is a reference point of traders' reactions. As it often happens, traders would regard the 52-week high as a resistance and refuse to bid the price higher even though some really good news arrives for the stock and eventually creates a continuation pattern. A similar scenario for bad news hitting the market can be analyzed as well.

Note that the 52-week highs and lows are publicly known information, and indeed, they are reported daily on some of the prints. Therefore, the 52-week high momentum strategy poses a strong challenge to the efficient-market hypothesis. And what is more, the 52-week high momentum strategy barely needs any computation, unlike some other trading indicators which involves calculations of the historical data.

In Marshall and Cahan (2005), the 52-week high momentum strategy is tested for Australian stocks, the first out-of-sample test outside the US. The reason is explained in Marshall and Cahan (2005) as the actively traded Australian stocks receive relatively little academic attention. The results in Marshall and Cahan (2005) show that the 52-week high momentum strategy still beats the other two momentum investment strategies in Jegadeesh and Titman (1993); Moskowitz and Grinblatt (1999) in the Australian stock market from 1991 to 2003.

12.3 Intraday and Interday Momentum Strategies

There are other momentum strategies apart from the 52-week high momentum strategy, e.g., one that involves intraday and interday information [Lam *et al.* (2007); Tian and Guo (2007)]. The intraday and interday momentum strategies apply the opening and closing prices, unlike most of the traditional technical indicators.

Intraday Momentum Strategy

For the intraday momentum strategy, first an indicator known as the M-period average intraday momentum (AIM) at time j is defined as

$$AIM_{M,j} = \frac{1}{M} \sum_{k=0}^{M-1} \left| x_{j-k}^c - x_{j-k}^o \right|,$$

where x_j^o and x_j^c are the opening and closing prices at time j respectively. A basic trading strategy based on the AIM is to buy at time j when

$$x_j^c > x_j^o \quad \text{and} \quad x_j^c - x_j^o > AIM_{M,j} \times p; \tag{12.3}$$

and sell at time j when

$$x_j^o > x_j^c \quad \text{and} \quad x_j^o - x_j^c > AIM_{M,j} \times p, \tag{12.4}$$

where $p \geq 1$ is a parameter. A further trading strategy based on (12.3) and (12.4) requires a certain surge or plummet in stock price's behavior. It is to buy at time j when

$$(12.3) \quad \text{and} \quad x_j^c - x_j^o > x_j^h - x_j^c + x_j^o - x_j^l;$$

and sell at time j when

$$(12.4) \quad \text{and} \quad x_j^o - x_j^c > x_j^h - x_j^o + x_j^c - x_j^l,$$

where x_j^h and x_j^l are the intraday high and intraday low at time j respectively. Note that these two intraday momentum strategies are closely related to the Japanese candlesticks. The terms $x_j^c - x_j^o$, $x_j^h - x_j^c$ and $x_j^o - x_j^l$, are the body, upper and lower shadows of a white candle respectively. On the other hand, the terms $x_j^o - x_j^c$, $x_j^h - x_j^o$ and $x_j^c - x_j^l$ are the counterparts for a black candle; see Section 2.1. The more advanced trading strategy requires an additional condition that the body is larger than the upper and lower shadows combined.

Interday Momentum Strategy

Similarly, for the interday momentum strategy, first an indicator known as the M-period average interday momentum (AOM) at time j is defined as

$$AOM_{M,j} = \frac{1}{M} \sum_{k=0}^{M-1} \left| x_{j-k}^c - x_{j-k-1}^c \right|,$$

where two consecutive closing prices are used. Then, we can design a basic interday momentum strategy as follows. We buy at time j when

$$x_j^c > x_{j-1}^c \quad \text{and} \quad x_j^c - x_{j-1}^c > AOM_{M,j} \times p; \tag{12.5}$$

and sell at time j when

$$x_{j-1}^c > x_j^c \quad \text{and} \quad x_{j-1}^c - x_j^c > AOM_{M,j} \times p, \tag{12.6}$$

where p is defined as before in the intraday momentum strategy.

Combined strategies using the intraday and interday momentum strategies can also be defined. An investor buys at time j when (12.3) or (12.5) is satisfied, and sells at time j when (12.4) or (12.6) is satisfied. An alternative for the combined strategy is to buy at time j when (12.3) and (12.5) are both satisfied, and sell at time j when (12.4) and (12.6) are both satisfied.

The intraday and interday momentum strategies (including the combined ones) are compared with the buy-and-hold strategy in Lam *et al.* (2007). The momentum strategies outperform the buy-and-hold strategy for most Asian indices, especially in the Shanghai and Shenzhen markets. As a result, the intraday and interday momentum strategies are suitable for traders who would like to take into account the surge and plummet of the stock prices during the day.

12.4 Relative Strength Index Indicators

In technical analysis, the relative strength index (RSI) is an indicator that shows the strength or weakness in price information. The RSI is proposed by J. Welles Wilder in Wilder (1978). Given a set of data points $\{x_j\}$, we first define, at time j, the variables of upward change and downward change:

$$u_j = \begin{cases} x_j - x_{j-1}, & \text{if } x_j > x_{j-1}, \\ 0, & \text{otherwise,} \end{cases}$$

and

$$d_j = \begin{cases} x_{j-1} - x_j, & \text{if } x_j < x_{j-1}, \\ 0, & \text{otherwise.} \end{cases}$$

At any time j, both u_j are d_j are nonnegative numbers and at least one of them is zero. When time j's closing price is strictly bigger than time $j-1$'s closing price, i.e., $x_j > x_{j-1}$, then u_j is positive and equals to the rise $x_j - x_{j-1}$, and hence represents the upward change. On the other hand, when time j's closing price is strictly smaller than time $j-1$'s closing price, i.e., $x_j < x_{j-1}$, then d_j is positive and equals to the drop $x_{j-1} - x_j$, and hence represents the downward change. If the price stays steady, i.e., $x_j = x_{j-1}$ for some j, which is not very likely to happen for liquid instruments in a real stock market, then we have $u_j = d_j = 0$.

The value of RSI at time j is defined as

$$RSI_j = 100 \times \frac{\sum_{k=0}^{\infty} \left(\frac{M-1}{M+1}\right)^k u_{j-k}}{\sum_{k=0}^{\infty} \left(\frac{M-1}{M+1}\right)^k u_{j-k} + \sum_{k=0}^{\infty} \left(\frac{M-1}{M+1}\right)^k d_{j-k}},$$

where the M-period exponential moving averages (EMAs) of the upward change and downward change are used in the fraction; see (5.5). Note that $\frac{2}{M+1}$ is a common factor and hence gets cancelled out in the fractional format. It is obvious that RSI_j is a number between 0 and 100. When the upward changes have an average close to zero, then the RSI will approach 0. When the downward changes have an average close to zero, then the RSI will approach 100. Wilder (1978) proposes to use the averaging with period $M = 27$ ($\alpha = \frac{1}{14}$) in the formulas.

Different trading strategies based on the RSI are studied in Wong *et al.* (2003). First, one needs to specify a lower bound L and an upper bound U

for the RSI range. For instance, the lower and upper bounds for the original Wilder-defined RSI are $L = 30$ and $U = 70$ respectively. The bound levels are also related to the period M. For instance, an RSI with longer period is better used with $L = 40$ and $U = 60$ [Wong *et al.* (2003)].

The first trading method using RSI depends on the arguments that when the RSI touches the lower and upper bounds L and U, the market is oversold and overbrought respectively. Hence, an investor will buy when the RSI crosses L, and sell when the RSI crosses U. The second trading method is to buy when the RSI has crossed L and then reverses its direction, and sell when the RSI has crossed U and then reverses its direction. The third trading method is to buy when the RSI has crossed L and then retraces all the way back to L, and sell when the RSI has crossed U and then retraces all the way back to U. The fourth trading method does not make use of the lower and upper bounds L and U. One simply buys when the RSI rises above 50, and sells when the RSI drops below 50.

In Wong *et al.* (2003), it is argued that the trading rules involving the RSI lead to mixed results in Singapore's stock market from 1974 to 1994. The reason is that the RSI-based rules are applied in the whole time period while the market is sometimes trending and sometimes not. However, the RSI is more effective during non-trending cycles. Therefore, one needs to determine the trending strength in the market before applying RSI-based technical indicators. The average directional movement index (ADX) is such a trend-detecting tool and will be introduced in Section 12.5.

12.5 Directional Indicators

The profitability of directional indicators is studied by Lam and Chong (2006). The set of direction indicators is developed by J. Welles Wilder in Wilder (1978) and it is the fundamental part of defining the ADX, which can be used as a trend detector before applying certain technical indicators. In this section, we introduce how directional indicators are used to construct a trading method in Lam and Chong (2006).

Directional indicators not only make use of the daily closing price, but also the intraday high and low prices. Assume that the high, low, and closing prices at time j are denoted by x_j^h, x_j^l, and x_j^c respectively. To define the directional indicators, we first define the directional movements. Essentially, directional movements depict whether intraday high and low prices follow an upward or downward evolution. The positive directional movement (DM$^+$) has the following value at time j:

$$DM_j^+ = \begin{cases} x_j^h - x_{j-1}^h, & \text{if } x_j^h > x_{j-1}^h \text{ and } x_j^h - x_{j-1}^h > x_{j-1}^l - x_j^l, \\ 0, & \text{otherwise.} \end{cases}$$

Note that $DM_j^+ \geq 0$ at any time j. If $DM_j^+ > 0$ at time j, it implies that time j's intraday high x_j^h must be higher than time $j-1$'s intraday high x_{j-1}^h. However, the inverse is not necessarily true, i.e., $x_j^h > x_{j-1}^h$ does not infer that $DM_j^+ > 0$. If a drop in neighbouring intraday low prices $x_{j-1}^l - x_j^l$ is larger than a rise in neighbouring intraday high prices $x_j^h - x_{j-1}^h$, then $DM_j^+ = 0$ at time j. All in all, positive directional movement DM$^+$ describes the upward movement in intraday high prices given that the downward movement in intraday low prices is relatively weaker. On the other hand, the negative directional movement (DM$^-$) has the following value at time j:

$$DM_j^- = \begin{cases} x_{j-1}^l - x_j^l, & \text{if } x_{j-1}^l > x_j^l \text{ and } x_{j-1}^l - x_j^l > x_j^h - x_{j-1}^h, \\ 0, & \text{otherwise.} \end{cases}$$

One can state similar arguments as the positive directional movements and conclude that DM$^-$ describes the downward movement in intraday low prices given that the upward movement in intraday high prices is relatively weaker. At any time j, DM_j^+ are DM_j^- are nonnegative and at least one of them equals to zero. Note that DM_j^+ are DM_j^- can be both zero at time j, e.g., when time j's price range is shorter than time $j-1$'s one such that $x_j^h < x_{j-1}^h$ and $x_{j-1}^l < x_j^l$.

For the next step, we calculate the true range (TR). TR includes the closing price in the previous day and hence measures the true trading range [Wilder (1978)]. The TR value at time j is

$$TR_j = \max\left\{x_j^h - x_j^l, x_j^h - x_{j-1}^c, x_{j-1}^c - x_j^l\right\},$$

or equivalently

$$\begin{aligned} TR_j &= \max\left\{x_j^h - x_j^l, \left|x_j^h - x_{j-1}^c\right|, \left|x_{j-1}^c - x_j^l\right|\right\} \\ &= \max\left\{x_j^h, x_{j-1}^c\right\} - \min\left\{x_j^l, x_{j-1}^c\right\}. \end{aligned}$$

One can verify the equivalence using the fact that $x_j^h > x_j^l$ at any time j.

The M-period positive directional indicator (DI$_M^+$) and the M-period negative directional indicator (DI$_M^-$) are defined next. Using DM_j^+, DM_j^-, and TR_j at time j, we define DI_M^+ at time j as the M-period EMA of DM^+ divided by the M-period EMA of TR:

$$DI_{M,j}^+ = 100 \times \frac{\sum_{k=0}^{\infty} \left(\frac{M-1}{M+1}\right)^k DM_{j-k}^+}{\sum_{k=0}^{\infty} \left(\frac{M-1}{M+1}\right)^k TR_{j-k}}.$$

Note that the factor $\frac{2}{M+1}$ that appears in front of the summation in M-period EMAs is cancelled in this fractional format; see (5.5). Similarly, the DI_M^- at time j is defined as the M-period EMA of DM^- divided by the M-period EMA of TR:

$$DI_{M,j}^- = 100 \times \frac{\sum_{k=0}^{\infty} \left(\frac{M-1}{M+1}\right)^k DM_{j-k}^-}{\sum_{k=0}^{\infty} \left(\frac{M-1}{M+1}\right)^k TR_{j-k}}.$$

Next, the M-period directional movement index (DX_M) at time j is defined as

$$DX_{M,j} = \left| \frac{DI_{M,j}^+ - DI_{M,j}^-}{DI_{M,j}^+ + DI_{M,j}^-} \right|.$$

Lastly, the ADX [Wilder (1978)] is defined as 100 times the M-period EMA of DX_M:

$$ADX_{M,j} = 100 \times \frac{2}{M+1} \sum_{k=0}^{\infty} \left(\frac{M-1}{M+1}\right)^k DX_{M,j-k}.$$

In the original framework, the period is selected as $M = 27$ ($\alpha = \frac{1}{14}$) [Wilder (1978)].

The ADX has been discussed in Section 11.3 for its trend detecting capability. Here, we focus on how directional indicators are used as technical indicators. In Lam and Chong (2006), the simple trading method is to monitor the crossovers between the M-period positive and negative directional indicators. At time j, a trading day is a buy day if

$$DI_{M,j}^+ > DI_{M,j}^-,$$

and a sell day if

$$DI^+_{M,j} < DI^-_{M,j}.$$

More trading methods can be created using the simple one together with more buy/sell refinement; see [Lam and Chong (2006)] for details. We also remark that the directional indicators therein use the SMA instead of the EMA. Lam and Chong (2006) test the directional indicators and estimate the returns in various stock markets. They find out that the directional indicators with $M = 10$, 14, and 20 outperform the buy-and-hold strategy in the stock markets of Hong Kong, South Korea, Japan, and Taiwan. For the UK and US stock markets, comparatively fewer directional indicators perform well, possibly owing to the market efficiency in those regions.

Appendix A

Relevant Literature in Different Topics

Classical Technical Analysis

There is an ocean of books or monographs that focus on classic technical analysis ideas such as chart pattern, trend lines, support and resistance levels. Readers can find the relevant information in Achelis (2000); Edwards *et al.* (2007); Kirkpatrick and Dahlquist (2010); Murphy (1999); Pring (2002), to name a few. More specifically, the point and figure chart is discussed in du Plessis (2012). Furthermore, the ancient but mysteriously useful Japanese candlesticks are also summarized and analyzed in Nison (1994, 2001). The more advanced *ichimoku kinko hyo*, or the Japanese cloud chart, is studied in the monograph [Linton (2010)].

A wide range of chart patterns can be found in the exhaustive book [Bulkowski (2005)]. For classic theories, Dow theory is explained in a modern manner in the definite work [Schannep (2008)]. On the other hand, the Elliott wave principle originates from Elliott (1938), and the classic book [Frost and Prechter (2005)] covers this topic in details. For various trading cycles, one can find resources from Appel (2005) written by MACD's creator Gerald Appel. Fibonacci trading is a branch in technical analysis that makes use of the Fibonacci ratios, and the materials are discussed in Boroden (2008). Gann lines and retracements can be found in the original document [Gann (1935)], or [Reddy (2012)] which focuses on Gann's trading methods.

Technical Market Indicators

Major technical indicators, like moving averages, can be found in general technical analysis books like Ehlers (2001); Lee and Tryde (2012); Mak (2003, 2006); Meyers (2011); Pring (2002). In particular, the encyclopedic book [Colby (2002)] contains an enormous list of technical indicators. Gerald Appel, the developer of the famous MACD technique has a book on technical analysis [Appel (2005)], so has John Bollinger, the creator of Bollinger bands [Bollinger (2001)]. The relative strength index (RSI) and average directional movement index (ADX) are proposed by J. Welles Wilder. The framework for these two widely used indicators can be found in Wilder's 1978 book [Wilder (1978)].

Regarding the theory behind linear filters like moving averages, digital signal processing books like Oppenheim and Schafer (1989); Damelin and Miller (2012) or the ones with MATLAB examples [Ingle and Proakis (2000); Stearns and Hush (2011)] are good references as well. Statistical analysis plays an important part in analyzing the profitability of technical indicators, and the influential paper [Brock *et al.* (1992)] provides an extensive study on the Dow Jones Index from 1897 to 1986. Readers are also referred to the overall review of technical analysis' performance from 1960 to 2004 in Park and Irwin (2007).

Wavelets and Empirical Mode Decomposition

The wavelet analysis is a mature subject in signal and image processing and [Mallet (2008)] is the representative text that guides newcomers from the very beginning. There are also books like Damelin and Miller (2012) that starts from basic digital signal processing and then gradually moves onto wavelets. One also does not want to miss Daubechies' ten lectures on wavelets [Daubechies (1992)]. On the other hand, there are wavelet books like [Fugal (2009); Weeks (2010)] which provide explanatory MATLAB examples, while [Gençay *et al.* (2001)] provides financial and economical applications.

The Hilbert–Huang transform with empirical mode decomposition (EMD) is a modern method for analyzing nonlinear and non-stationary signals, and it was first introduced by Norden E. Huang in the seminal paper [Huang *et al.* (1998)]. In Huang *et al.* (2003), EMD together with Hilbert spectral analysis is applied in finance, e.g., to define a form of time-variant volatility. Since EMD is just an algorithm, it is bound to computational problems. EMD-related issues and a MATLAB package are available in Rilling *et al.* (2003, 2007). Meanwhile, there is an alternative version of EMD in MATLAB codes made public by Norden E. Huang's team:

`http://rcada.ncu.edu.tw/research1.htm`

Appendix B

Descriptive Statistics for Tested Markets

In Chapters 7, 8, and 9, we have measured the profitability of various technical trading rules in Hong Kong's and Chinese stock markets from 2007 to 2011. In this part, we report the descriptive statistics for the two tested markets during this period. Table B.1 contains the summary statistics for the entire period for one-day returns on the Hong Kong and Chinese stock markets. The descriptive statistics include the distribution characteristics: mean, variance, skewness, and kurtosis. Moreover, the first five autocorrelations, the Bartlett standard error for the autocorrelations, and the Ljung–Box Q statistics at the fifth lag are also reported.

In Table B.1, the distribution for the Hong Kong stock market is lightly right-skewed, and the distribution for the Chinese stock market shows negative skewness. The returns in both markets are moderately leptokurtic for

Table B.1: Summary statistics for daily returns in the whole period from 2007 to 2011 for Hong Kong's and Chinese stock markets

	Hong Kong	**China**
N	1233	1215
Mean	−7.8592e−5	−0.00017355
Std.	0.020658	0.020248
Skewness	0.089221	−0.35983
Kurtosis	8.7949	5.3273
$\rho(1)$	−0.035764	−0.0051022
$\rho(2)$	0.016051	−0.017498
$\rho(3)$	−0.055201	0.034818
$\rho(4)$	−0.037377	0.05931
$\rho(5)$	−0.0056982	−0.017597
Bartlett Std.	0.028479	0.028689
$Q(5)$	7.4433	6.5574

the entire period. Most of the autocorrelations are generally small and insignificant, expect for $\rho(4)$, the fourth order serial correlation in the Chinese stock market, which is greater than two times the Bartlett asymptotic standard error and hence significant. Note that autocorrelations should be insignificant if the market is efficient because new information arrives in a random way. The Ljung–Box portmanteau test statistics are not high, owing to the fact that both samples have a relatively short period of 5 years.

Appendix C

General Remarks on Wavelet-based and EMD-based Trading Rules

Note that we have used the 10 moving average crossover rules according to Brock *et al.* (1992) to prevent selection bias. It is because there is a possibility that one moving average crossover rule will work extremely well if enough combinations are tested. One might then be confused by the wavelet-based and the EMD-based indicators because there are four parameters j_1, k_1, j_2, and k_2 involved. As a result, these two methods are seemingly more affected by selection bias. We remark that the lag numbers j_1 and j_2 are related to k_1 and k_2 respectively. For instance, in the EMD-based trading scheme, when the number of IMF summation changes, its instantaneous frequency also changes and hence the lag number is different. However, in our studies, we simply chose j_1 and j_2 as some common round numbers like 5, 50, and 100. Determining j_1 and j_2 from k_1 and k_2 respectively can be an interesting topic. But once that is done, then the wavelet-based and EMD-based indicators will only have two changing parameters k_1 and k_2, which have much fewer combinations than thousands of moving average crossover rules. In Tables C.1 and C.2, we report some additional test results on Hong Kong's and Chinese stock markets using more moving average crossover rules and EMD-based trading rules with new parameters.

Table C.1: Additional standard test results and break-even costs for Hang Seng Index from 2007 to 2011, with moving average crossover rules and EMD-based trading rules

Rule	N_b	N_s	$\hat{\mu}_b$ (T_b)	$\hat{\mu}_s$ (T_s)	$\hat{\mu}_b - \hat{\mu}_s$ (T_{bs})	C
(1, 50, 2%)	495	439	0.00046 (0.56)	−0.00077 (−0.48)	0.00123 (0.81)	0.0069
(5, 50, 0%)	645	588	0.00015 (0.26)	−0.00033 (−0.21)	0.00047 (0.39)	0.0075
(5, 50, 1%)	574	508	0.00014 (0.24)	−0.00042 (−0.26)	0.00056 (0.42)	0.0070
(5, 50, 2%)	483	427	0.00024 (0.33)	−0.00033 (−0.17)	0.00057 (0.37)	0.0066
(1, 150, 2%)	591	501	0.00068 (0.89)	−0.00078 (−0.52)	0.00146 (1.08)	0.0209
(5, 150, 2%)	591	503	0.00046 (0.62)	−0.00102 (−0.71)	0.00148 (1.10)	0.0392
(1, 200, 2%)	632	470	0.00027 (0.42)	−0.00060 (−0.38)	0.00088 (0.63)	0.0127
(2, 200, 2%)	633	473	0.00042 (0.60)	−0.00046 (−0.28)	0.00088 (0.64)	0.0202
(5, 200, 0%)	705	528	0.00028 (0.44)	−0.00056 (−0.37)	0.00084 (0.66)	0.0351
(5, 200, 1%)	666	500	0.00031 (0.46)	−0.00068 (−0.45)	0.00099 (0.74)	0.0341
(5, 200, 2%)	631	476	0.00027 (0.41)	−0.00065 (−0.42)	0.00093 (0.68)	0.0302
(1, 0, 4, 50, 2%)	567	474	0.00035 (0.48)	−0.00052 (−0.33)	0.00087 (0.63)	0.0099
(2, 5, 4, 50, 2%)	567	484	0.00048 (0.61)	−0.00090 (−0.62)	0.00138 (0.99)	0.0202
(1, 0, 5, 100, 2%)	581	532	0.00071 (0.88)	−0.00055 (−0.38)	0.00126 (0.98)	0.0220
(2, 5, 5, 100, 2%)	585	544	0.00103 (1.19)	−0.00108 (−0.83)	0.00211 (1.65)	0.0566
(3, 15, 5, 100, 2%)	586	539	0.00082 (0.97)	−0.00091 (−0.68)	0.00173 (1.35)	0.0512

Table C.2: Additional standard test results and break-even costs for China Shanghai Composite Index from 2007 to 2011, with moving average crossover rules and EMD-based trading rules

Rule	N_b	N_s	$\hat{\mu}_b$ (T_b)	$\hat{\mu}_s$ (T_s)	$\hat{\mu}_b - \hat{\mu}_s$ (T_{bs})	C
(1, 50, 2%)	507	521	0.00156 (1.72)	−0.00119 (−0.88)	0.00275 (2.13)	0.0214
(5, 50, 0%)	585	630	0.00087 (1.07)	−0.00115 (−0.95)	0.00202 (1.75)	0.0385
(5, 50, 1%)	550	572	0.00113 (1.31)	−0.00151 (−1.23)	0.00264 (2.16)	0.0494
(5, 50, 2%)	508	503	0.00136 (1.49)	−0.00145 (−1.09)	0.00281 (2.12)	0.0458
(1, 150, 2%)	595	529	0.00107 (1.27)	−0.00169 (−1.37)	0.00276 (2.24)	0.0511
(5, 150, 2%)	603	521	0.00094 (1.14)	−0.00178 (−1.44)	0.00272 (2.20)	0.0832
(1, 200, 2%)	573	543	0.00101 (1.18)	−0.00147 (−1.17)	0.00248 (2.00)	0.0458
(2, 200, 2%)	570	540	0.00094 (1.11)	−0.00155 (−1.24)	0.00249 (2.00)	0.0571
(5, 200, 0%)	630	585	0.00083 (1.05)	−0.00126 (−1.03)	0.00209 (1.79)	0.0786
(5, 200, 1%)	606	560	0.00108 (1.29)	−0.00146 (−1.19)	0.00255 (2.13)	0.1136
(5, 200, 2%)	572	534	0.00121 (1.38)	−0.00164 (−1.32)	0.00285 (2.28)	0.1205
(1, 0, 4, 50, 2%)	543	598	0.00166 (1.83)	−0.00173 (−1.48)	0.00339 (2.84)	0.0745
(2, 5, 4, 50, 2%)	536	580	0.00190 (2.05)	−0.00177 (−1.50)	0.00367 (3.01)	0.0890
(1, 0, 5, 100, 2%)	582	521	0.00123 (1.40)	−0.00197 (−1.58)	0.00319 (2.52)	0.0511
(2, 5, 5, 100, 2%)	596	521	0.00134 (1.53)	−0.00164 (−1.31)	0.00298 (2.39)	0.0719
(3, 15, 5, 100, 2%)	603	519	0.00150 (1.70)	−0.00184 (−1.49)	0.00334 (2.69)	0.0930

Appendix D

MATLAB Functions Used in Context

MATLAB is used for all the data charting and numerical experiments throughout this book. In this appendix, we summarize the MATLAB functions used in the context.

Aside from the special price charts like bar chart, point and figure chart, and Japanese candlesticks, all the other price charts are produced by the MATLAB function `plot`. The easiest way is to type in

```
plot(data)
```

where `data` is a vector containing the stock prices downloaded from Yahoo! Finance. Then, we can modify the raw figure by using the MATLAB plot tools and add more features like graph legends, descriptive text and etc. The MATLAB function `subplot` is also used to put related figures together.

Special Price Charts

- Figure 2.1

The bar chart is produced by the MATLAB function `highlow` (which is known as the high, low, open, close chart in MATLAB). It is a built-in function contained in the Financial Toolbox. In order to produce such a chart, we need to specify four vectors `high`, `low`, `close`, `open`, which correspond to the daily's highs, lows, closing prices and opening prices of the data. Then, the following will do the work:

```
highlow(high,low,close,open)
```

- Figure 2.3

The point and figure chart is produced by the MATLAB function `pointfig`, which is another built-in function from the Financial Toolbox. The following will produce a point and figure chart with input stock prices contained in a vector `data`:

```
pointfig(data)
```

- Figure 2.4

The Japanese candlesticks can also be created by the built-in function `candle` from the Financial Toolbox. Since the candlestick chart is another version of bar chart, it also requires four data inputs `high`, `low`, `close`, `open` corresponding to daily's highs, lows, closing prices and opening prices:

```
candle(high,low,close,open)
```

Simple Linear Filters and Frequency Response

The MATLAB function `tsmovavg` is contained in the Financial Toolbox and it is used to derive moving averages. For example, the following

```
tsmovavg(data,'s',20)
```

returns a vector containing the simple moving average values with duration chosen as $M = 20$ given the `data` vector. Here, `'s'` specifies that a simple moving average is used. Alternatively,

```
tsmovavg(data,'e',20)
```

will return the exponential moving average with $\alpha = 2/(20+1)$.

For the frequency responses that appear in Chapters 5 and 6, we decide not to use MATLAB build-in functions from the Signal Processing Toolbox because we want to illustrate how the DTFT is related to the derivation of the frequency response. Here, the MATLAB command `fft` is frequently used to create all those frequency response figures.

- Figure 5.3

```
M = 10;
N = 1024;
omega = 0:2*pi/(N-1):2*pi;
H = (1+2*sum(cos(kron((1:M).',omega)),1))/(2*M+1);
plot(omega,abs(H))
```

- Figure 5.4

```
M = 20;
N = 1024;
omega = 0:2*pi/(N-1):2*pi;
num = ones(1,M)/M;
H = fft(num,N);
subplot(2,1,1)
plot(omega,abs(H))
subplot(2,1,2)
plot(omega,angle(H))
```

Note that the variable N stands for the sample nodes in an N-point DTFT and it is chosen as the power of two (e.g., 1024) since the built-in function `fft` in MATLAB works faster with such a number.

- Figure 5.6

```
M = 20;
alpha = 2/(M+1);
N = 1024;
omega = 0:2*pi/(N-1):2*pi;
num = alpha;
den = [1,alpha-1];
H = fft(num,N)./fft(den,N);
subplot(2,1,1)
plot(omega,abs(H))
subplot(2,1,2)
plot(omega,angle(H))
```

In Section 5.4, we have mentioned that the convolution of two linear filters would return the coefficients of the combined filter. This can be verified by the following MATLAB codes that would return the filter coefficients:

```
h1 = [1/2,1/2];
h2 = [1/3,1/3,1/3];
conv(h1,h2)
```

For the convolution theorem, suppose we get back to the example in Section 5.4 with the first filter being a 2-period simple moving average and the second filter being a 3-period simple moving average. The following two sets of MATLAB codes:

```
h1 = [1/2,1/2];
h2 = [1/3,1/3,1/3];
fft(conv(h1,h2),8)
```

and

```
h1 = [1/2,1/2];
h2 = [1/3,1/3,1/3];
fft(h1,8).*fft(h2,8)
```

will return the same result and hence confirm the results stated by the convolution theorem.

- Figure 5.7

```
M1 = 10; M2 = 20;
alpha1 = 2/(M1+1); alpha2 = 2/(M2+1);
N = 1024;
omega = 0:2*pi/(N-1):2*pi;
num1 = alpha1; num2 = alpha2;
den1 = [1,alpha1-1]; den2 = [1,alpha2-1];
H1 = fft(num1,N)./fft(den1,N);
H2 = fft(num2,N)./fft(den2,N);
H = H1.*H2;
subplot(2,1,1)
plot(omega,abs(H1),omega,abs(H2),omega,abs(H))
subplot(2,1,2)
plot(omega,angle(H1),omega,angle(H2),omega,angle(H))
```

- Figure 6.1

```
N = 1024;
omega = 0:2*pi/(N-1):2*pi;
H = (1-exp(-i*omega))/2;
subplot(2,1,1)
plot(omega,abs(H))
subplot(2,1,2)
plot(omega,angle(H))
```

- Figure 6.2

```
M = 20;
N = 1024;
omega = 0:2*pi/(N-1):2*pi;
num = ones(1,M)/M;
H = ones(1,N) - fft(num,N);
subplot(2,1,1)
plot(omega,abs(H))
subplot(2,1,2)
plot(omega,angle(H))
```

- Figure 6.4

```
M1 = 10; M2 = 50;
alpha1 = 2/(M1+1); alpha2 = 2/(M2+1);
N = 1024;
omega = 0:2*pi/(N-1):2*pi;
num1 = alpha1; num2 = alpha2;
den1 = [1,alpha1-1]; den2 = [1,alpha2-1];
H1 = fft(num1,N)./fft(den1,N);
H2 = fft(num2,N)./fft(den2,N);
H = H1 - H2;
subplot(2,1,1)
plot(omega,abs(H))
subplot(2,1,2)
plot(omega,angle(H))
```

- Figure 6.6

```
N = 1024;
omega = 0:2*pi/(N-1):2*pi;
num = [11/6,-3,3/2,-1/3];
H = fft(num,N);
subplot(2,1,1)
plot(omega,abs(H))
subplot(2,1,2)
plot(omega,angle(H))
```

- Figure 6.8

```
M1 = 12; M2 = 26; M3 = 9;
alpha1 = 2/(M1+1); alpha2 = 2/(M2+1);
alpha3 = 2/(M3+1);
N = 1024;
omega = 0:2*pi/(N-1):2*pi;
num1 = alpha1; num2 = alpha2; num3 = alpha3;
den1 = [1,alpha1-1]; den2 = [1,alpha2-1];
den3 = [1,alpha3-1];
H1 = fft(num1,N)./fft(den1,N);
H2 = fft(num2,N)./fft(den2,N);
H3 = fft(num3,N)./fft(den3,N);
H = (H1 - H2).*(ones(1,N) - H3);
subplot(2,1,1)
plot(omega,abs(H))
subplot(2,1,2)
plot(omega,angle(H))
```

- Figure 6.9

Figure 6.9 with the MACD and its signal line is produced by the Financial Toolbox function macd:

```
[macdvec, nineperma] = macd(data)
```

It returns the MACD line and its signal line, contained in the vector macdvec and nineperma respectively, given the input vector data filled with stock prices.

Wavelets and Empirical Mode Decomposition

We have only made use of the multilevel wavelet decomposition in our studies. For instance, the following commands contained in MATLAB's Wavelet Toolbox will return a 3-level wavelet decomposition using a order 2 Daubechies wavelet:

```
[C,L] = wavedec(data,3,'db2');
d1 = wrcoef('d',C,L,'db2',1);
d2 = wrcoef('d',C,L,'db2',2);
d3 = wrcoef('d',C,L,'db2',3);
r = wrcoef('a',C,L,'db2',3).
```

Here, the input data will be decomposed as the sum of d1, d2, d3, and r, which respectively stand for the wavelet details and the residual mean, denoted as $\mathbf{d}_1$, $\mathbf{d}_2$, $\mathbf{d}_3$, and $\mathbf{r}$ in Section 8.2.

The build-in function wavedec belongs to the Wavelet Toolbox. It performs a multilevel one-dimensional wavelet analysis using a specified wavelet type (like 'db2' in the above routine) and a number of levels (like 3 in the above routine). Then, another built-in function wrcoef from the Wavelet Toolbox could be used to reconstruct the signal components with the outputs C and L from wavedec. The parameters 'd' and 'a' determine whether detail or approximation coefficients are reconstructed.

In Chapter 9, we use the empirical mode decomposition to create a trading strategy. In our numerical experiments, we simply apply the existing EMD algorithm in MATLAB format designed by Rilling *et al.* (2007). For instance, the following

```
imf = emd(data);
```

will return all the intrinsic mode functions together with a residual contained in a matrix imf, given the inputs contained in data. Each row of the matrix imf corresponds to one intrinsic mode function, except for the last row. For instance, the first row imf(1,:) is the first intrinsic mode function, which is first removed from the original signal and contains the highest-frequency components. The last row `imf(end,:)` represents the trend of the input, which itself is not an intrinsic mode function; see Section 9.2.

Gann's Trend Lines and Retracements

In Chapter 10, we have discussed the Gann's trend lines and retracements, and their roles in forming trend, resistance or support. Suppose we now have the input as `data`. The following MATLAB codes will generate a figure containing Gann's trend lines with the significant angles: $82.5, 75, 71.25, 63.75, 45, 26.25, 18.75, 15,$ and 7.5, and Gann's retracements for the following levels: $\frac{1}{8}, \frac{2}{8}, \frac{3}{8}, \frac{4}{8}, \frac{5}{8}, \frac{6}{8},$ and $\frac{7}{8}$. The only parameter to be determined is the variable `ratio`, which is the ratio between the price unit and the time unit. Since the default time unit is 1 (one day) in this case, the ratio simply represents the price unit. For instance, with `ratio = 2`, one uses a price unit of 2 points and a time unit of one day.

```
ratio = 2;
N = length(data);
[ymin,xmin] = min(data);
[ymax,xmax] = max(data);

%Gann's retracements
Grtrc = kron(((1:7)*ymin + (7:-1:1)*ymax)/8,ones(N,1));

%Gann's significant angles
deg = [82.5,75,71.25,63.75,45,26.25,18.75,15,7.5];
deg = tan(deg*pi/180)*ratio;

%Gann's upward trend lines
Guln = ymin + kron(deg,(1:N).' - xmin);
Guln(1:xmin-1,:)= NaN;

%Gann's downward trend lines
Gdln = ymax - kron(fliplr(deg),(1:N).' - xmax);
Gdln(1:xmax-1,:)= NaN;

plot(1:N,[data,Grtrc,Guln,Gdln])
```

Bollinger Bands and Relative Strength Index

In Chapters 11 and 12, we have implemented the ideas of Bollinger bands and relative strength index (RSI) for market entry timings. Given the input `data`, the built-in function `bollinger` from the Financial Toolbox

```
[mid, uppr, lowr] = bollinger(data);
```

returns three vectors `mid`, `uppr` and `lowr` corresponding to the 20-period simple moving average, the upper and lower bands which are two standard deviations away. Note that the duration of the moving average and the number of standard deviations of the bands can be modified by giving more input parameters in `bollinger`.

Given the input `data`, the built-in function `rsindex` from the Financial Toolbox

```
rsi = rsindex(data);
```

returns a vector `rsi` containing the RSI values. The default number of periods is set as 14. Note that this parameter can be modified by having one more input parameter in the function `rsindex`.

Bibliography

Achelis, S. (2000). *Technical Analysis from A to Z* (McGraw-Hill, New York).

Allen, H. and Taylor, M. P. (1989). Charts and fundamentals in the foreign exchange market, Bank of England discussion paper no. 40.

Allen, H. and Taylor, M. P. (1992). The use of technical analysis in the foreign exchange market, *Journal of International Money and Finance* **11**, 3, pp. 304–314.

Appel, G. (2005). *Technical Analysis: Power Tools for Active Investors* (FT Press, Upper Saddle River, New Jersey).

Ariff, M. and Wong, W.-K. (1996). Risk-premium effect on share market prices: A test of loanable funds prediction, RMIT Economics and Finance Seminar Paper.

Ball, R. (1978). Anomalies in relationships between securities' yields and yield-surrogates, *Journal of Financial Economics* **6**, 2–3, pp. 103–126.

Balvers, R. J., Cosimano, T. F. and McDonald, B. (1990). Predicting stock returns in an efficient market, *The Journal of Finance* **45**, 4, pp. 1109–1128.

Breen, W., Glosten, L. R. and Jagannathan, R. (1989). Economic significance of predictable variations in stock index returns, *The Journal of Finance* **44**, 5, pp. 1177–1189.

Bessembinder, H. and Chan, K. (1995). The profitability of technical trading rules in Asian stock markets, *Pacific-Basin Finance Journal* **3**, 2–3, pp. 257–284.

Bollinger, J. A. (2001). *Bollinger on Bollinger Bands* (McGraw-Hill, New York).

Boroden, C. (2008). *Fibonacci Trading: How to Master the Time and Price Advantage* (McGraw-Hill, New York).

Brealey, R. A. and Myers, S. C. (1991). *Principles of Corporate Finance* (McGraw-Hill, New York).

Brock, W., Lakonishok, J. and LeBaron, B. (1992). Simple technical trading rules and the stochastic properties of stock returns, *The Journal of Finance* **47**, 5, pp. 1731–1764.

Bulkowski, T. N. (2005). *Encyclopedia of Chart Patterns*, 2nd edn. (Wiley Trading, New Jersey).

Campbell, J. Y. (1987). Stock returns and the term structure, *Journal of Financial Economics* **18**, 2, pp. 373–399.

Campbell, J. Y. and Shiller, R. J. (1988a). The dividend–price ratio and expectations of future dividends and discount factors, *Review of Financial Studies* **1**, 3, pp. 195–228.

Campbell, J. Y. and Shiller, R. J. (1988b). Stock prices, earnings, and expected dividends, *The Journal of Finance* **43**, 3, pp. 661–676.

Carter, R. B. and Van Auken, H. (1990). Securities analysis and portfolio management: A survey and analysis, *Journal of Portfolio Management* **16**, 3, pp. 81–85.

Chew, B.-K., Tan, W. T. and Wong, W.-K. (1996). Unraveling mysticism in Gann's theory: Prophecy of local stock market trends, *Singapore Journal of Stock Exchange*, February, pp. 14–18.

Cheung, Y.-W. and Wong, C. Y.-P. (1999). Foreign Exchange Traders in Hong Kong, Japan, and Singapore: A Survey Study, in *Advances in Pacific Basin Financial Markets* (Theodore Bos and Thomas A Fetherston eds.), Volume V, pp. 111–134.

Cheung, W., Lam, K. S. K and Yeung, H. (2011). Intertemporal profitability and the stability of technical analysis: Evidences from the Hong Kong stock exchange, *Applied Economics* **43**, 15, pp. 1945–1963.

Cochrane, J. H. (1991). Production-based asset pricing and the link between stock returns and economic fluctuations, *The Journal of Finance* **46**, 1, pp. 209–238.

Colby, R. W. (2002). *The Encyclopedia of Technical Market Indicators*, 2nd edn. (McGraw-Hill, New York).

Conrad, J. and Kaul, G. (1988). Time-variation in expected returns, *Journal of Business* **61**, 4, pp. 409–425.

Curcio, R. and Goodhart, C. (1991). The clustering of bid/ask prices and the spread in the foreign exchange market, London School of Economics, Financial Markets Group Discussion Paper 110.

Damelin, S. B. and Miller, W. Jr. (2012). *The Mathematics of Signal Processing* (Cambridge University Press, Cambridge, England).

Daubechies, I. (1992). Ten lectures on wavelets, *SIAM*, 1992.

DeBondt, W. F. M. and Thaler, R. H. (1985). Does the stock market overreact? *The Journal of Finance* **40**, 3, pp. 793–805.

DeBondt, W. F. M. and Thaler, R. H. (1987). Further evidence on investor overreaction and stock market seasonality, *The Journal of Finance* **42**, 3, pp. 557–581.

De Long, J. B., Shleifer, A., Summers, L. H. and Waldmann, R. J. (1990). Positive-feedback investment strategies and destabilizing rational speculation, *The Journal of Finance* **45**, 2, pp. 379–395.

Drakakis, K. (2008). Empirical mode decomposition of financial data, *International Mathematical Forum* **3**, 25, pp. 1191–1202.

du Plessis, J. (2012). *The Definitive Guide to Point and Figure: A Comprehensive Guide to the Theory and Practical Use of the Point and Figure Charting Method*, 2nd edn. (Harriman House, Petersfield, UK).

Edwards, R. D., Magee, J. and Bassetti, W. H. C. (2007). *Technical Analysis of Stock Trends*, 9th edn. (AMACOM, New York).

Ehlers, J. F. (2001). *Rocket Science for Traders: Digital Signal Processing Applications* (Wiley Trading, New Jersey).

Elliott, R. N. (1938). *The Wave Principle.* Republished (1980/1994): R. N. *Elliott's Masterworks.* Robert R. Prechter, Jr., ed. Gainesville, GA: New Classics Library, p. 144.

Fama, E. F. (1965). The behavior of stock-market prices, *Journal of Business* **38**, 1, pp. 34–105.

Fama, E. F. (1970). Efficient capital markets: A review of theory and empirical work, *The Journal of Finance* **25**, 2, pp. 383–423.

Fama, E. F. and Blume, M. E. (1966). Filter rules and stock-market trading, *Journal of Business* **39**, 1, pp. 226–241.

Fama, E. F. and French, K. R. (1988). Permanent and temporary components of stock prices, *Journal of Political Economy* **96**, 2, pp. 246–273.

Fama, E. F. and French, K. R. (1989). Business conditions and expected returns on stocks and bonds, *Journal of Financial Economics* **25**, 1, pp. 23–49.

Frankel, J. A. and Froot, K. A. (1986). Understanding the U.S. dollar in the eighties: The expectations of chartists and fundamentalists, *The Economic Record*, pp. 24–38.

Frankel, J. A. and Froot, K. A. (1990a). The rationality of the foreign exchange rate. Chartists, fundamentalists and trading in the foreign exchange market, *American Economic Review* **80**, 2, pp. 181–185.

Frankel, J. A. and Froot, K. A. (1990b). Chartists, Fundamentalists and the Demand for Dollars, in *Private Behaviour and Government Policy in Interdependent Economies* (Oxford University Press, New York), pp. 73–126.

Friedman, M. (1953). The Case for Flexible Exchange Rate, in *Essays in Positive Economics* (University of Chicago Press, Chicago).

Froot, K. A., Scharfstein, D. S. and Stein, J. C. (1992). Herd on the street: Informational inefficiencies in a market with short-term speculation, *The Journal of Finance* **47**, 4, pp. 1461–1484.

Frost, A. J. and Prechter, R. R. Jr. (2005). *Elliott Wave Principle: Key To Market Behavior*, 10th edn. (New Classics Library, Gainesville, GA).

Fugal, D. L. (2009). *Conceptual Wavelets in Digital Signal Processing* (Space & Signals Technical Publishing, San Diego).

Gann, W. D. (1935). The basis of my forecasting method.

Gençay, R., Selçuk, F. and Whitcher, B. (2001). *An Introduction to Wavelets and Other Filtering Methods in Finance and Economics* (Academic Press, Waltham, Massachusetts, USA).

George, T. J. and Hwang, C.-Y. (2004). The 52-week high and momentum investing, *The Journal of Finance* **59**, 5, pp. 2145–2176.

Grossman, S. J. (1976). On the efficiency of competitive stock markets where traders have diverse information, *The Journal of Finance* **31**, 2, pp. 573–584.

Grossman, S. J. and Stiglitz, J. (1976). Information and competitive price systems, *American Economic Review* **66**, 2, pp. 246–253.

Guhathakurta, K., Mukherjee, I. and Chowdhury, A. R. (2008). Empirical mode decomposition analysis of two different financial time series and their comparison, *Chaos, Solitons and Fractals* **37**, 4, pp. 1214–1227.

Haar, A. (1910). Zur Theorie der orthogonalen Funktionensysteme, *Mathematische Annalen* **69**, pp. 331–371.

Hong, L. (2011). Decomposition and forecast for financial time series with high-frequency based on empirical mode decomposition, *Energy Procedia* **5**, pp. 1333–1340.

Huang, N. E. *et al.* (1998). The empirical mode decomposition and the Hilbert spectrum for nonlinear and non-stationary time series analysis, *Proceedings of the Royal Society A* **454**, 1971, pp. 903–995.

Huang, N. E. *et al.* (2003). Applications of Hilbert–Huang transform to non-stationary financial time series analysis, *Applied Stochastic Models in Business and Industry* **19**, 3, pp. 245–268.

Ingle, V. K. and Proakis, J. G. (2000). *Digital Signal Processing Using MATLAB* (Brooks/Cole, Pacific Grove, CA).

Jegadeesh, N. and Titman, S. (1993). Returns to buying winners and selling losers: Implications for stock market efficiency, *The Journal of Finance*, **48**, 1, pp. 65–91.

Jensen, M. C. and Bennington, G. A. (1970). Random walks and technical theories: Some additional evidence, *The Journal of Finance* **25**, 2, pp. 469–482.

Kirkpatrick, C. D. and Dahlquist, J. (2010). *Technical Analysis: The Complete Resource for Financial Market Technicians*, 2nd edn. (FT Press, Upper Saddle River, New Jersey).

Kitchin, J. (1923). Cycles and trends in economic factors *Review of Economic Statistics* **5**, 1, pp. 10–16.

Kondratiev, N. D. (1925). *The Major Economic Cycles (in Russian)* (Moscow). Translated and published as *The Long Wave Cycle* (Richardson & Snyder, New York).

Kung, J. J. and Wong, W.-K. (2009). Efficiency of the Taiwan stock market, *The Japanese Economic Review* **60**, 3, pp. 389–394.

Kung, J. J., Carverhill, A. P. and McLeod, R. H. (2010). Indonesia's stock market: Evolving role, growing efficiency, *Bulletin of Indonesian Economic Studies* **46**, 3, pp. 329–346.

Kwon, K.-Y. and Kish, R. J. (2002). Technical trading strategies and return predictability: NYSE, *Applied Financial Economics* **12**, 9, pp. 639–653.

Lam, W.-S. and Chong, T. T.-L. (2006). Profitability of the directional indicators, *Applied Financial Economics Letters* **2**, 6, pp. 401–406.

Lam, V. W.-S., Chong, T. T.-L. and Wong, W.-K. (2007). Profitability of intraday and interday momentum strategies, *Applied Economics Letter* **14**, 15, pp. 1103–1108.

Lee, R. and Tryde, P. (2012). *Timing Solutions for Swing Traders: A Novel Approach to Successful Trading Using Technical Analysis and Financial Astrology* (Wiley Trading, New Jersey).

Levich, R. M. and Thomas III, L. R. (1993). The signicance of technical trading-rule prots in the foreign exchange market: A bootstrap approach, *Journal of International Money and Finance* **12**, 5, pp. 451–474.

Lin, L., Wang, Y. and Zhou, H. (2009). Iterative filtering as an alternative algorithm for empirical mode decomposition, *Advances in Adaptive Data Analysis* **1**, 4, pp. 543–560.

Linton, D. B. (2010). *Cloud Charts: Trading Success with the Ichimoku Technique* (Updata plc, London).

Lo, A. W. and MacKinlay, A. G. (1990). When are contrarian profits due to stock market overreaction? *Review of Financial Studies* **3**, 2, pp. 175–205.

Lui, Y.-H. and Mole, D. (1998). The use of fundamental and technical analyses by foreign exchange dealers: Hong Kong evidence, *Journal of International Money and Finance* **17**, 3, pp. 535–545.

Mak, D. K. (2003). *The Science of Financial Market Trading* (World Scientific, Singapore).

Mak, D. K. (2006). *Mathematical Techniques in Financial Market Trading* (World Scientific, Singapore).

Mallet, S. (1989). A theory for multiresolution signal decomposition: The wavelet representation, *IEEE Transactions on Pattern Analysis and Machine Intelligence* **11**, 7, pp. 674–693.

Mallet, S. (2008). *A Wavelet Tour of Signal Processing: The Sparse Way*, 3rd edn. (Academic Press, San Diego).

Marshall, B. R. and Cahan, R. M. (2005). Is the 52-week high momentum strategy profitable outside the US? *Applied Financial Economics* **15**, 18, pp. 1259–1267.

Metghalchi, M., Chang, Y.-H. and Marcucci, J. (2008). Is the Swedish stock market efficient? Evidence from some simple trading rules, *International Review of Financial Analysis* **17**, 3, pp. 475–490.

Meyers, T. (2011). *The Technical Analysis Course: Learn How to Forecast and Time the Market*, 4th edn. (McGraw-Hill, New York).

Mills, T. C. (1997). Technical analysis and the London Stock Exchange: Testing trading rules using the FT30, *International Journal of Finance and Economics* **2**, 4, pp. 319–331.

Mitra, S. K. (2011). How rewarding is technical analysis in the Indian stock market? *Quantitative Finance* **11**, 2, pp. 287–297.

Moskowitz, T. J. and Grinblatt, M. (1999). Do industries explain momentum? *The Journal of Finance*, **54**, 4, pp. 1249–1290.

Murphy, J. J. (1999). *Technical Analysis of the Financial Markets: A Comprehensive Guide to Trading Methods and Applications* (New York Institute of Finance, Manhattan, New York).

Nison, S. (1994). *Beyond Candlesticks: New Japanese Charting Techniques Revealed* (John Wiley & Sons, New Jersey).

Nison, S. (2001). *Japanese Candlestick Charting Techniques* (Prentice Hall Press, 2nd edn., New Jersey).

Oppenheim, A. V. and Schafer, R. W. (1989). *Discrete-time Signal Processing* (Prentice Hall, Englewood Cliffs, New Jersey).

Park, C.-H. and Irwin, S. H. (2007). What do we know about the profitability of technical analysis? *Journal of Economic Surveys* **21**, 4, pp. 786–826.

Pring, M. J. (2002). *Technical Analysis Explained: The Successful Investor's Guide to Spotting Investment Trends and Turning Points*, 2nd edn. (McGraw-Hill, New York).

Reddy, H. (2012). *The Trading Methodologies of W. D. Gann: A Guide to Building Your Technical Analysis Toolbox* (FT Press, Upper Saddle River, New Jersey).

Renshaw, E. (1993). Modeling the stock market for forecasting purposes, *Journal of Portfolio Management* **20**, 1, pp. 76–81.

Rilling, G., Flandrin, P. and Gonçalvès, P. (2003). On empirical mode decomposition and its algorithms, *IEEE-EURASIP Workshop Nonlinear Signal Image Processing (NSIP)*, Grado, Italy, June 8–11, 2003.

Rilling, G., Flandrin, P. and Gonçalvès, P. (2007). http://perso.ens-lyon.fr/patrick.flandrin/emd.html.

Samuelson, P. (1989). The judgment of economic science on rational portfolio management, *Journal of Portfolio Management* **16**, 1, pp. 4–12.

Schannep, J. (2008). *Dow Theory for the 21st Century: Technical Indicators for Improving Your Investment Results* (Wiley, New Jersey).

Schulmeister, S. (1988). Currency speculation and dollar fluctuations, Banca Nazionale del Lavoro Quartely Review, December 1988, pp. 343–365.

Sharpe, W. F. (1975). Adjusting for risk in performance measurement, *Journal of Portfolio Management* **1**, 2, pp. 29–34.

Shiller, R. J. (1984). Stock prices and social dynamics, *Brooking Papers on Economic Activity* 2, pp. 457–498.

Shiller, R. J. (1987). Investor behavior in the October 1987 stock market crash: Survey evidence, NBER Working Paper No. 2446.

Stearns, S. D. and Hush, D. R. (2011). *Digital Signal Processing with Examples in MATLAB*, 2nd edn. (CRC Press, Boca Raton, Florida).

Sweeney, R. J. (1986). Beating the foreign exchange market, *The Journal of Finance* **41**, 1, pp. 163–182.

Sy, W. (1990). Market timing: Is it a folly? *Journal of Portfolio Management* **16**, 4, pp. 11–16.

Tian, G. G., Wan, G. H. and Guo, M. (2002). Market efficiency and the returns to simple technical trading rules: New evidence from U.S. equity market and Chinese equity markets, *Asia-Pacific Financial Markets* **9**, 3, pp. 241–258.

Tian, G. G. and Guo, M. (2007). Interday and intraday volatility: Additional evidence from the Shanghai Stock Exchange, *Review of Quantitative Finance and Accounting* **28**, 3, pp. 287–306.

Weeks, M. (2010). *Digital Signal Processing Using MATLAB & Wavelets*, 2nd edn. (Jones and Bartlett Publishers, Burlington, Massachusetts).

Wilder, J. W. (1978). *New Concepts in Technical Trading Systems* (Trend Research).

Wong, M. C.-S. and Wong, K. F. (1993). Are simple market timing skills useful in the Hong Kong stock market? In *Asia-Pacific Financial and Forecasting Research Centre, City Polytechnic of Hong Kong.*

Wong, W.-K. (1993). An equity indicator for the Singapore market, *Journal of Singapore Stock Exchange*, May, pp. 21–23.

Wong, W.-K. (1994). A composite equity indicator for Singapore, *Journal of Singapore Stock Exchange*, May, pp. 24–26.

Wong, M. C.-S. (1995). Market reactions to several popular trend-chasing technical signals, *Applied Economics Letters* **2**, 11, pp. 449–456.

Wong, M. C.-S. (1997). Fund management performance, trend-chasing technical analysis and investment horizons: A case study, *Omega* **25**, 1, pp. 57–63.

Wong, W.-K., Chew, B.-K. and Sikorski, D. (2002). Can the forecasts generated from E/P ratio and bond yield be used to beat stock markets? National University of Singapore, Department of Economics in its series Departmental Working Papers with number wp0201.

Wong, W.-K., Manzur, M. and Chew, B.-K. (2003). How rewarding is technical analysis? Evidence from Singapore stock market, *Applied Financial Economics* **13**, 7, pp. 543–551.

Wong, W.-K., Du, J. and Chong, T. T.-L. (2005). Do the technical indicators reward chartists? A study on the stock markets of China, Hong Kong and Taiwan, *Review of Applied Economics* **1**, 2, pp. 183–205.

Wong, W.-K. and McAleer, M. (2009). Mapping the presidential election cycle in US stock markets, *Mathematics and Computers in Simulation* **79**, 11, pp. 3267–3277.

Vandell, R. F. and Stevens, J. L. (1989). Evidence of superior performance from timing, *Journal of Portfolio Management* **15**, 3, pp. 38–42.

Vasiliou, D., Eriotis, N. and Papathanasiou, S. (2006). How rewarding is technical analysis? Evidence from Athens Stock Exchange, *Operational Research* **6**, 2, pp. 85–102.

Zhu, T. (2006). Suspicious financial transaction detection based on empirical mode decomposition method, *IEEE Asia-Pacific Conference on Services Computing (APSCC)*, pp. 300–304.

Index

Printed in the United States
By Bookmasters